Transistor Audio Amplifiers

by

Jack Darr

HOWARD W. SAMS & CO., INC.
THE BOBBS-MERRILL CO., INC.
INDIANAPOLIS • KANSAS CITY • NEW YORK

FIRST EDITION
FIRST PRINTING—1971

International Standard Book Number: 0-672-20838-5
Library of Congress Catalog Card Number: 70-149393

Preface

This is a book about a very popular item—something that you will see a lot of, if you haven't already. Transistor audio amplifiers come in all shapes and sizes, from almost invisible ones used in small phonographs, up to the monsters with enough power output to break windows, used by musical groups.

You're going to find an amazing similarity in these units. As a matter of fact, the 250- to 300-watt musical instrument amplifiers use *exactly* the same output circuit as some of the 50-milliwatt phonograph amplifiers! So, the only difference in troubleshooting methods, etc., will be the actual value of the voltage and current readings.

This is a book aimed at simplifying transistor circuits. Transistors have a great many parameters that can be tested, measured, and noted. Only about four of these have any usefulness to us. What we are interested in is finding a defective transistor and then finding a suitable replacement for it. A transistor is a very small but very complex device; granted, it takes a lot of complicated operations to make one, but we are not interested in that, either. Remember this: it is the *function* of the device, and not its detailed structure that matters!

Think of a transistor in terms of what it does, not what it is. It amplifies a signal; that's all. It does this by controlling the collector current with a small voltage on the base—exactly the same basic action as a vacuum tube controlling its plate current with a small voltage on the grid. So, when we have troubles, we look for a stage that is not amplifying the signal, and we've got it.

The original writers on transistors talked of current amplification, holes, minority carriers, barriers, and many other things. This succeeded in confusing the average technician (the writer among them) for some years. Transistors are not thoroughly understood, even today, by the majority of working technicians, simply because we (writers) haven't explained in simple terms how they work. Forget all of these details, and concentrate on only one thing. "Is this thing amplifying a *signal?*" Despite the current-amplifier confusion, we can still take test readings in the oldest way known—with a quick jab of a voltmeter probe, requiring no breaking of circuits, etc. If the voltage is normal, then the current will be normal IF the thing is working. We will make current measurements, of course, but no more often than we do in tube circuits.

The "secret," if there is one, of successful servicing of solid-state equipment is twofold: one, a thorough knowledge of how the circuit works—that is, a knowledge of the basic circuitry used—, and two, a knowledge of how it works when it isn't working. If you'd like that in more literate terms: the ability to select the right test equipment and apply it to the circuit properly to obtain the information you need about what's going on, and you've got it.

So, we'll start out with a full discussion of the basic transistor circuits that you'll find, then go on to actual circuitry used in the amplifiers that you'll be working on. Test equipment, test methods, and case histories of actual troubles will be found throughout this book. Every test and piece of equipment and test method mentioned has been "bench tested" in actual servicing of solid-state equipment over a period of some years.

This is a book aimed at giving the working technician a good practical knowledge of transistor circuits, as found in real equipment of all sizes. The only piece of new test equipment you'll need will be a good transistor tester. Your regular radio-tv test equipment will do anything else you need to do. No "high-precision" equipment is needed. We have worked out a simple way of finding replacements for any type of transistor, based on how it is being used in the actual equipment. You will find this helpful; we have found it so.

I hope that this book will be of some help to the technicians who have been confused by transistor servicing. Writing it has certainly helped me! Good luck.

JACK DARR

Contents

CHAPTER 5

CHAPTER 6

CHAPTER 7

CHAPTER 8

CHAPTER 9

CHAPTER 10

CHAPTER 1

Transistors: Characteristics, Typical Circuits

If you're going to be an expert transistor service technician, the first thing you need to learn about is—transistors. This sounds pretty obvious, but it is true. If you're an old hand with a good knowledge of vacuum-tube theory, electron flow, etc., it won't take long. If you're just getting started, it might even be easier. Before you can diagnose troubles in these circuits, you have to know how they work. Once you've mastered that, you can start in learning "how they work when they aren't working!" In other words, the typical *fault reactions* that you'll find in defective circuitry. To be able to recognize a fault reaction, you must know exactly what the *normal* reaction ought to be.

SOME FUNDAMENTAL POINTS ABOUT TRANSISTORS

Simpler Than Tubes

When you get right down to it, transistors are actually a good deal simpler than tubes. Almost all of them are simple triodes; tubes can have numerous elements, and can get pretty complicated. However, either of them can be pretty simple if you look at them in terms of "what they do," rather than "what they are." They both do exactly the same thing—amplify signals. A little later on, we'll show you a lot of easy tests for finding the place where the signal stops. THERE is where the trouble is—there is where you start your detailed tests.

Voltages and Polarity

There is one thing found in transistor circuits that is never found in tubes. The polarity of the applied voltage can be reversed. The plate of a tube is always positive. Transistors can be npn, with a positive voltage applied to the collector, or pnp, with a negative voltage on the collector. Here are a couple of facts that will help you remember what polarity you should find. Memorize them.

In a transistor, the polarity of applied voltage that causes current through a junction is called "forward bias." This is the same as the polarity of the element to which it is applied. Positive for "p" elements, negative for "n" elements. The type-name of the transistor will tell you which polarity goes where. In the expressions "pnp" and "npn," the *middle* letter tells you the polarity of the *base* material. In a schematic diagram, the arrow of the emitter always "points to negative." If it points inward, toward the base, the base is negative, and the transistor is a pnp. If it points outward, toward the emitter terminal, the base is positive, and the transistor is npn.

The base is the *controller element;* like the grid in a vacuum tube, it controls the current in the output circuit, plate, or collector.

Transistors have one peculiarity, when compared with tubes. A certain "minimum bias" must be applied to the base before there will be any collector current. This is 0.2 volt for a germanium transistor, and 0.6 volt for silicon transistors. Now, here's the fact:

All transistors must have this minimum bias applied to the base, or they will not work at all. If the transistor is at zero bias (same voltage on base and emitter), it is completely cut off, no current will flow.

By contrast, a tube with zero bias will usually draw a fairly heavy plate current; a strong negative bias must be applied to its control grid to make it cut off. Fact No. 2:

In a tube, the "current-increasing-bias"—the polarity that causes more current—is always positive. In transistors, you can have current-increasing bias of either polarity. It will always be the same polarity as that of the base material.

Third rule: All transistors, as used in the standard circuits, will have a reverse-bias on the *collector*, measured from the emitter. Npn transistors will thus always have a positive voltage on collectors; pnp transistors will have negative collector voltages. This will always be the highest voltage on the transistor, read from its own emitter. Fact No. 3:

Transistors always have a reverse bias on the collector.

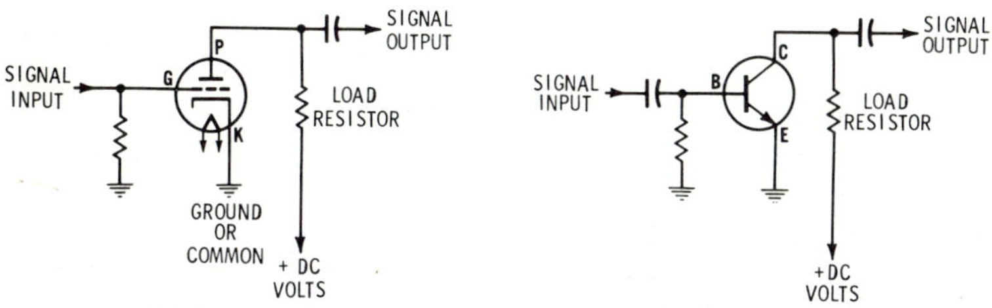

(A) Tube circuit. (B) Transistor circuit.

Fig. 1-1. Comparison of tube and transistor circuitry.

TRANSISTOR CIRCUITS

Now, let's look at the standard circuits that you will find in this type of transistor equipment. There aren't very many. Three stock circuits are used with almost all transistor amplifiers: common-emitter, common-collector, and common-base. The common-emitter circuit will be the one you'll see most often, so, let's take it first. There will be variations and modifications, and in some cases a very deceptive-looking imitation, but they'll all be the same. This corresponds to the common-cathode circuit of a tube.

We said "common-emitter," etc. By "common," we mean the element that is used in both the input and output circuits. In the common-emitter circuit, the input circuit is actually "base-to-emitter," and the output circuit is "collector-to-emitter." Fig. 1-1 shows a direct comparison between transistor and tube circuits. In such circuits, all voltages should be measured from the common element—the cathode in a tube, the emitter in this transistor.

There's a reason for this. In a tube, the only voltage that has any effect in attracting electrons from the cathode to the plate is the voltage *difference* between the plate and cathode. In a transistor, the voltage between emitter and collector is the only one that has any effect. (Never mind the "electrons or holes" bit; it's still the *voltage* that makes them move, and that's what we're really interested in!)

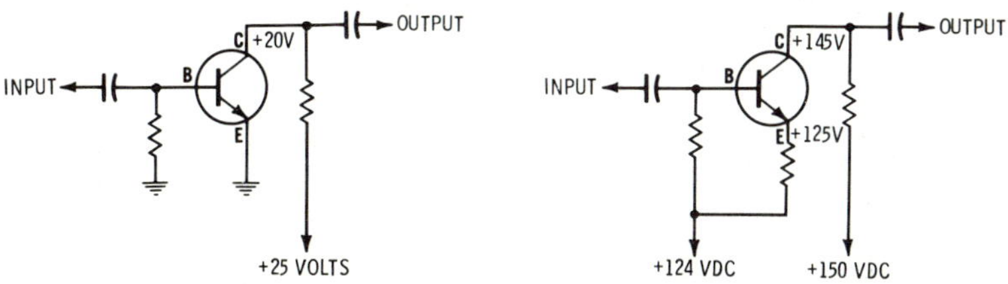

(A) Grounded emitter. (B) Emitter not grounded.

Fig. 1-2. Two circuits apparently different but with similar action.

Fig. 1-2 shows an illustration of this, as you will find it in many actual circuits. In the circuit of Fig. 1-2A, the emitter is grounded. So, this is at zero voltage to ground. The +25-volt supply is fed to the collector through the load resistor, and we read a +20-volt difference from emitter to collector. This is the voltage difference that causes signal current. (In fact, you'll see here that there is a 5-volt drop across the load resistor, indicating that there *is* current through it. Remember this. We're going to use it over and over again.)

Now look at the circuit of Fig. 1-2B. Looks different? If we read the dc voltages from ground, yes. *But*, if we read the voltages directly on the transistor elements, we'll find that they are exactly the same as at A! The transistor still has a +20-volt emitter-to-collector voltage. There is still a 5-volt drop across the load resistor. So, the current will be the same, the gain of the stage will be the same, and so on. You will find this kind of circuit in a lot of actual amplifiers.

Circuit Variations

Now, let's look at a few of the more common variations of this circuit. Fig. 1-3 shows one that you're going to run into in a lot of different places. This is called a "grounded-collector" circuit. Despite the fact that the collector is ground, this is *not* a common-collector circuit. This may start an argument, but if you'll notice, the signal path through this transistor is exactly the same as in the previous ones. The signal is applied to the base and taken off at the *collector*. Also, the transistor dc voltages are exactly what they were.

The collector is grounded through a load resistor. A high negative voltage is fed to the emitter and base, through resistors. When a signal is applied, electrons flow from the supply, through the dropping resistor, through the emitter-collector junction, through the load resistor, and back to ground. The voltage-drop across the collector load resistor is the amplified signal voltage. (By the way, all "current-arrows" in drawings shown in this book will indicate the direction of *electron flow*, as in the collector-current arrowed line of Fig. 1-3.)

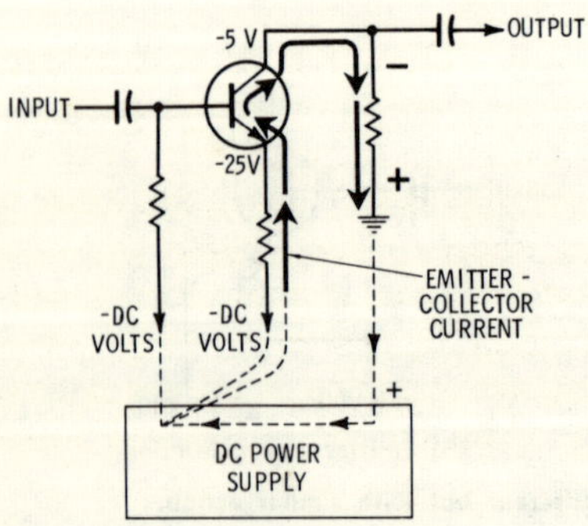

Fig. 1-3. Grounded-collector circuit.

The voltage drop across the collector load resistor is the same; 5 volts. Since the lower end of the load resistor is ground, this comes out as a −5 volts. So, a −25 volts is applied to the emitter. If we take a voltage reading from the emitter to the collector, what do we get? A +20 volt collector-to-emitter voltage, just as before (−5 volts is "more positive" than −25 volts).

The Common-Collector (Emitter-Follower)

The circuit shown in Fig. 1-4 *is* different. This is a common-collector circuit. Notice that the collector goes directly to the source of voltage. The input signal is applied to the base, just as before, but this time it is taken off at the emitter. This circuit is exactly the same as the cathode-follower used with tubes, and has the same characteristics.

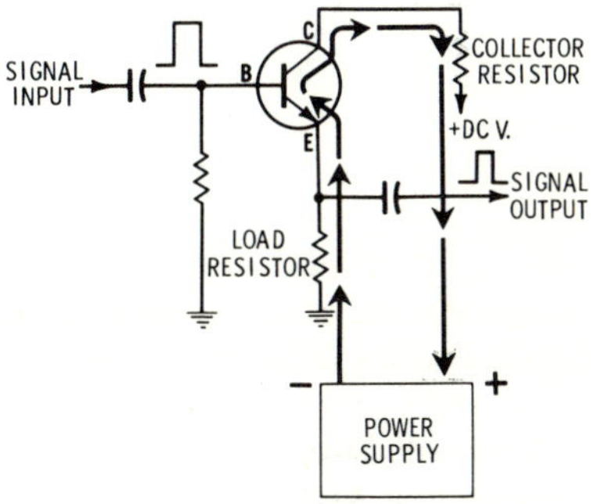

Fig. 1-4. Common-collector or emitter-follower.

In the common-emitter circuit, the signal polarity is inverted as it passes through; a positive-going pulse on the base comes out as a negative-going pulse on the collector. Here, we have no inversion; the base and emitter voltages are in phase. Also, there is seldom any voltage *gain* in these stages. These are used for such things as phase-inverting, impedance matching, and so on.

How does it work? Same as before. Follow the dc current path from the negative terminal of the power supply, through the emitter resistor, to emitter-collector junction, through the collector resistor (which is just a voltage-dropping resistor here), and back to the positive terminal of the power supply. Once again, the same *signal-current* flows in the same circuit. Thus, it goes through both the emitter resistor and collector resistor. It creates a signal voltage drop across each one. In this circuit, we're taking it off the emitter resistor, but remember that voltage drop on the collector resistor. We'll use it very soon.

The Common-Base Circuit

Fig. 1-5 shows a common-base circuit, mostly for the novelty. You'll see very few, if any, of these in audio-amplifier circuits. They're found mostly in rf-amplifier circuitry because of their special characteristics. The signal is fed into the emitter, and taken off at the collector. The base is grounded, and common to input and output circuits.

This type of circuit has a very low input impedance, but a high output impedance; just the opposite of what we usually want in audio work. Current gain is small, but the voltage gain can be made fairly high. So, if you do run across one of these, it will probably be used for impedance-matching. Voltage relations across the transistor elements will be the same as before.

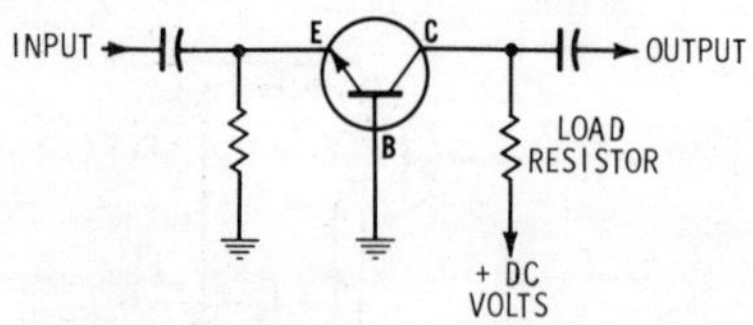

Fig. 1-5. Common-base circuit.

THE SIGNAL PATH

Up to this point we have dealt only with the dc voltage and current paths. You might call these the "static" conditions–the dc voltages that you'd find if you measured them with no signal input. (These happen to be the voltages given on all PHOTOFACT schematics, which are the only ones giving us a "standard" set of readings.)

Now, let's put this transistor to work and get some good out of it. Its work is to amplify signals. Just as we do with tubes, we should think of the ac signal and dc voltages as *separate* things. This makes circuits much easier to analyze. Actually, of course, we all know that they are the same. The current that flows in a circuit is a direct current modulated by an ac component. But, it's a lot easier if we can think of them as being two separate things. As a matter of fact, in all amplifying circuits, there are two separate paths for the signal and the dc supply. When we're analyzing a circuit, we should follow the signal-path until we find out where it stops; at that point, we start checking the dc path and find out why *it* has interrupted the signal.

Let's follow the signal-path through a typical amplifier stage. Fig. 1-6 shows a standard common-emitter circuit. The ac signal comes in *through* the coupling capacitor, to the base. (This is the major difference between the signal path and the dc paths. A capacitor is a short circuit for the ac signal, but a completely open circuit for

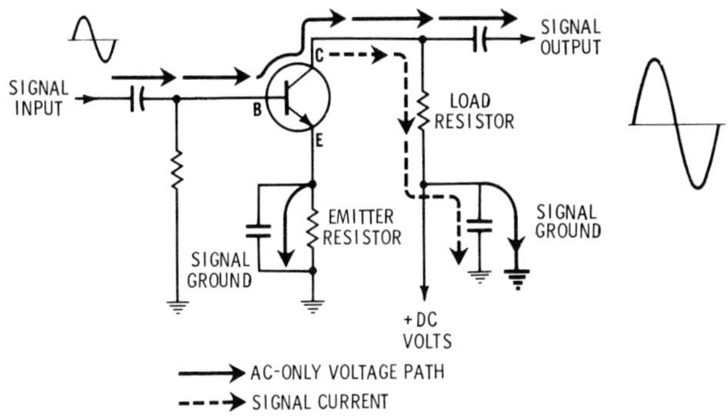

Fig. 1-6. Signal path through common-emitter stage.

dc.) On the base, by transistor action, a change in signal current causes a much larger change in the collector-emitter current through the load resistor. This ac voltage drop across the load resistor is the output signal. It has been amplified by the gain of the transistor.

The output signal (voltage) is taken off through another coupling capacitor and fed to the base of the following stage. This is the signal-path through all amplifier stages. To find out if it's working normally all we have to do is read the signal voltage amplitude at the input, then read the same signal at the output. Say we have a 1.0 volt peak-to-peak voltage on the base, and the stage has a voltage gain of 25. Then, we should read a 25-volt peak-to-peak signal on the collector, if the stage is working.

VOLTAGE AMPLIFICATION VERSUS CURRENT AMPLIFICATION

You have all probably read books that say that all transistors are "current amplifiers," and have become slightly confused. This is true, of course; a change in the base-emitter current can cause a change in the collector-emitter current. However, current measurements are far more difficult than voltage measurements, which require only a quick jab of a voltmeter probe. For some reason, "current-thinking" seems to be harder than "voltage-thinking."

We still use voltage amplification in a great many circuits, and we always will. Even in transistor circuits, there will always be *resistors* in the current paths. If so, we can always use the voltage drop that develops across these resistors to tell exactly how much current there is. Current and voltage in any circuit are always directly related to one another, and always will be until someone finds a way to repeal Ohm's law.

So, if you want to measure current in a given circuit, find a known resistor and read the voltage drop across it. There's an old Latin motto, "*Essentia non sunt multiplicanda praeter necessitatem,*" which just means "Do it the easy way." Voltage readings, because of their simplicity, have always been the favorite with working technicians, and always will be. Speed with accuracy are our prime concerns, and this is the fastest way.

To check any transistor circuit, take the voltage readings, then compare them with the ones shown on the schematic. If they match, fine; this circuit is almost certainly working normally. We go on until we find one that isn't. Any circuit defect that causes incorrect *current* will *always* cause the associated voltages to "go off" in direct ratio. You'll find, as we go along, that one of the things we look for, in most cases, is the transistor that is not conducting current at all; it is cut off. This is very easy to spot, by the incorrect voltages that you can read on it. Example: A transistor that is supplied through a dropping resistor from the dc power supply. Normal voltage on the collector is given as, say, +25 volts, from a supply of +35 volts. If you read +35 volts on the collector, you know instantly that this transistor isn't drawing *any* current! There is no normal voltage drop across the load resistor. So, you investigate; you have found *some* trouble, at least.

DEGENERATION AND FEEDBACK: BYPASSING

Up to now, we've shown only a few of the parts used in each circuit, mostly the resistors. There is another very important part: bypass capacitors. Since these affect only the signal path (ac), they show up for the first time in the circuit of Fig. 1-6. Let's see what they do.

The emitter bypass is an ac ground for signal frequencies. Its reactance is so very low for ac that the signal currents pass right on through it with practically no loss at all. By flowing through the capacitor, they bypass the fixed resistor in the emitter circuit. So, only the dc component of the emitter current will affect the base-emitter voltage. Since this is the *bias* on the transistor, we want to hold it as steady as possible. The low reactance of the bypass capacitor should send the ac signal to ground so easily that with the scope we can read no signal at all on the emitter.

What would happen if this capacitor weren't there, in other words, if it were open? We have a certain signal voltage variation (ac) on the base; let's say 1.0 volt peak-to-peak sine wave. Without the bypass capacitor, this signal voltage would flow through the emitter resistor, and develop a signal-voltage drop across it. This voltage drop is exactly in-phase with the input signal on the base.

The bias that determines the conduction of the transistor, and thus its gain, is the base-emitter voltage. If the emitter voltage is fixed by bypassing the signal variation *around* it, then the base voltage has full control of the conduction, and the transistor has the proper gain. The collector current will then vary in step with the base voltage, developing a much larger current variation across the load resistor. The higher the base voltage, the higher the collector current, and the greater the gain of the stage.

If the emitter bypass capacitor is open, the *emitter* voltage will vary exactly in step with the *base* voltage. Both signals are in-phase, remember. So, if the base voltage rises by 0.1 volt, so does the emitter voltage, and the bias is unchanged. But what we want the base voltage to do is change with respect to the emitter voltage. (Instantaneously, that is.) If the emitter is held fixed and the base increases by 0.1 volt, the collector current will also increase and the signal is amplified. If base and emitter voltages go up and down exactly in step, they will always maintain the same voltage difference and there will, in effect, be no bias *change* on the base.

This type of action is called "feedback." In this case, it is "degenerative feedback," meaning a feedback that reduces the gain of an amplifier. If the two voltages rose and fell exactly in step and held the same value at every instant, we would have 100 percent degenerative feedback, and the stage would have no gain at all. In actual cases, this isn't exactly what happens, but it's awfully close! If the emitter bypass opens, you find the same signal voltage on base and emitter and very little gain through the stage.

Later on, you'll find circuits that make use of this effect. Deliberate degeneration is introduced into some stages in order to cancel out distortion and improve frequency response, help to hold down the chances of transistor runaway, etc. By making the emitter bypass capacitors smaller than usual, a partial degeneration can be introduced, which help the frequency response of a stage but always reduces the gain by some amount. We'll show you such a stage very soon.

The Collector Bypass

There's another very important bypass capacitor seen in the circuit of Fig. 1-6. This is the collector bypass, or, more correctly, "load-resistor bypass." When the collector is drawing a large amount of signal current, this current goes through the load resistor. There it creates a large *signal-voltage* drop. This appears at the top of the load resistor and is usually measured *to ground*. The signal voltage is fed through the coupling capacitor to the input of the next stage, which is also "referred to ground." So, regardless of the dc voltages present, our signal voltages are almost always referred to ground or chassis.

To develop the maximum signal voltage across the load resistor, the "bottom" of the resistor must be kept at a true ac ground potential for signals only. In the grounded-collector stage, you remember, the bottom of the load resistor was actually grounded. In this example, we can't ground it, because of the high dc voltage. So, we make an ac ground through a big capacitor.

What would happen if this capacitor were not there? The bottom of the load resistor would be quite a way "off ground," because there are usually quite a few other resistors between this point and ground, through the dc power supply, dropping resistors, etc. So, the total signal voltage would then be developed across all of these resistors, instead of across the load resistor alone. Only a certain percentage of the signal would show up across the load resistor, and this would mean a severe loss of gain in the stage.

REGENERATION

The loss of gain, although annoying, isn't the worst thing that results from the absence of this bypass capacitor. The dc power supply is common to all other circuits in the amplifier; it supplies them all with operating power. So, there is a "common connection" between the bottom end of our load resistor and all other load resistors, dropping resistors, etc., in the whole circuit.

So, without the bypass capacitor, a good-sized *signal voltage* would appear on the dc supply lines. This signal would be fed back through the power-supply network, and get into all of the other stages in some form. Somewhere in there, it will find a stage where the phase relationships of the normal signal and this unwanted visiting signal will be just exactly right to cause "regeneration" (tending to *cause* oscillation; degeneration tends to stop it). So, the whole circuit will burst into oscillation; this is one of the more common troubles that you'll find in case histories.

This can be hard to spot, unless you know just where and how to look for it. Audible oscillation is easy to identify; the whole amplifier "howls" when it's turned on, without any input signal at all. However, very severe distortion with a loss of gain can be caused by the amplifier going into ultrasonic oscillation; the thing is oscillating violently, causing heavy distortion, etc., but at a frequency that is so high you can't hear it! This can happen anywhere from 20,000 Hz on up. The resulting disturbance to the circuitry will upset the bias, cause heavy currents to flow where they shouldn't, and raise hob in general. There is only one instrument that will instantly identify this kind of trouble, and that is the oscilloscope. We'll tell you how to use this later on.

In practical circuits, you'll find the load bypass capacitor where it's drawn in Fig. 1-6, as a separate unit at the bottom of the load resistor. In many of them, a small resistor, perhaps 1000 ohms, will be used between this point and the main dc supply line. This is not so much for any voltage drop as it is for filtering. It gives better isolation between this point and other similar points through the rest of the amplifier, and helps to prevent undesired feedback.

In the simpler and cheaper amplifiers, there may be no separate load bypasses. Their place will be taken by the output filter capacitor of the dc power supply. This will be a very large electrolytic capacitor, which does dual duty; it smooths out the ripple in the power supply and also acts as a bypass for all of the load resistors, etc., connected to that point. The reactance of these huge capacitors, often up to 1000 μF or more in capacitance, is so low that the dc power supply will have an impedance of practically zero to any signals that might get into it. They will be shunted off to ground and eliminated. When these capacitors go bad, the impedance rises, signals appear, and you've got trouble.

DELIBERATE DEGENERATION—THE "BOOTSTRAP" CIRCUIT

In some cases, you'll find circuits that make use of a combination of these effects. This is done for special purposes, usually to raise the input impedance of a transistor preamplifier stage, correct the frequency response, etc. The standard common-emitter circuit tends to have a pretty low input impedance, a few hundred ohms at the most. This isn't right for high-impedance devices such as crystal microphones or stereo (ceramic) cartridges, or even high-Z guitar pickups. The efficiency drops far below acceptable levels, and the frequency response may be affected. These are all "voltage" devices, with very small currents.

Fig. 1-7 shows the input stage of a Magnavox transistor amplifier. Here, to get a very high input impedance, a special feedback circuit called a "bootstrap" is used. The emitter resistor of the preamplifier transistor is made very large (4700 ohms) compared to the values you'll find in other circuits. This holds the transistor current down

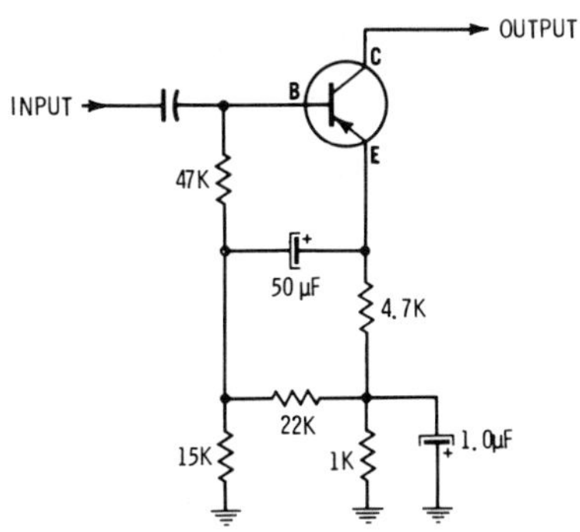

Fig. 1-7. Bootstrapped input circuit.

to a very low value, around 300 microamperes. Notice that there is no emitter bypass to ground. So, almost all of the base signal will appear across the 4700-ohm resistor. Now, it is *coupled* to the bottom of the base resistor, 47k, through the 50-μF capacitor.

Due to this coupling of the signal, the base resistor "sees" the same signal voltage at both top and bottom. There is no signal-voltage *difference* between them. So, this resistor looks like an *open circuit* to the input signal. If it were not for this effect, the input impedance of such a stage could never be greater than the sum of the resistors in the base circuit, that is $47k + 15k = 62k$ ohms. The unbypassed emitter resistor also makes the input impedance look bigger.

A rough rule of thumb for such circuits is to multiply the resistance in the emitter circuit by the gain of the transistor. Let's say that this one has a gain of 100. Then the total input impedance would be $100 \times 4700 = 470{,}000$ ohms. This will allow high-impedance devices such as crystal phono cartridges, etc., to develop a high signal voltage across the input. Due to the large amount of degenerative feedback in such a stage, its gain is not too great. So, once we've got the signal, we can follow it with a high-gain amplifier stage and get the signal up to the level needed to drive the output stage to full power. Degeneration improves the quality of the signal, but it does nothing at all for the gain. In testing such stages, be happy if you come out with as much as you put in. If you can "break even," fine.

COMMONLY USED BIAS CIRCUITS

We have said that all transistors must have a certain minimum bias applied before any current can flow. There are several different ways of getting this bias voltage. You'll find quite a few different circuits in commercial amplifiers. Learn to recognize each one, and how to test them. It's essential. Now let's look at these important circuits.

Fixed Bias

Fig. 1-8 shows a circuit used very often. This is called "fixed bias." Here, a resistor is connected from the high-voltage supply to the

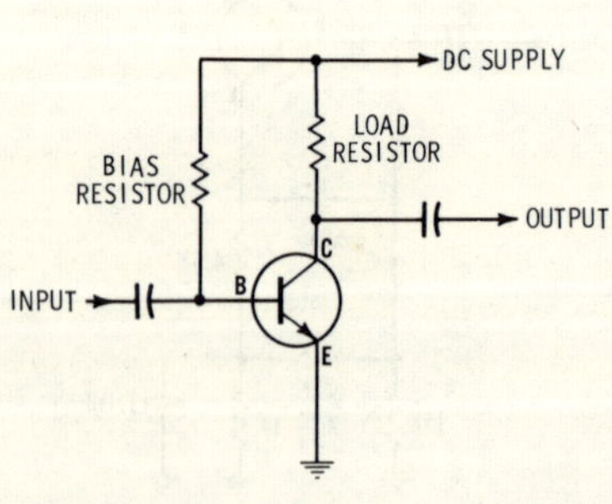

Fig. 1-8. Transistor fixed-bias circuit.

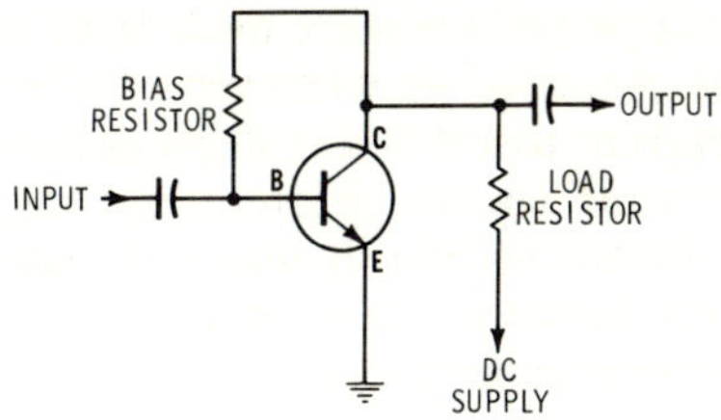

Fig. 1-9. Transistor self-bias circuit.

base. This resistor is large enough that the base-emitter current through it causes a voltage drop of just the right amount to set the base at the desired forward-bias voltage. This works, but it has its disadvantages. One of them is its sensitivity to small temperature changes. It can allow the transistor to go into thermal runaway and destroy itself. Therefore, you won't find a circuit this simple in common use.

Self Bias

Fig. 1-10 shows a popular circuit, one propably used more than any other. This uses a standard voltage divider with two resistors to the dc supply. The base current now flows through the collector load resistor. This causes some ac degeneration, reducing the possible gain. However, this degenerative action does cancel out the possibility of thermal runaway. It's also slightly more stable than the basic fixed-bias circuit. This circuit is found in a few more amplifiers than the original one, but not too many.

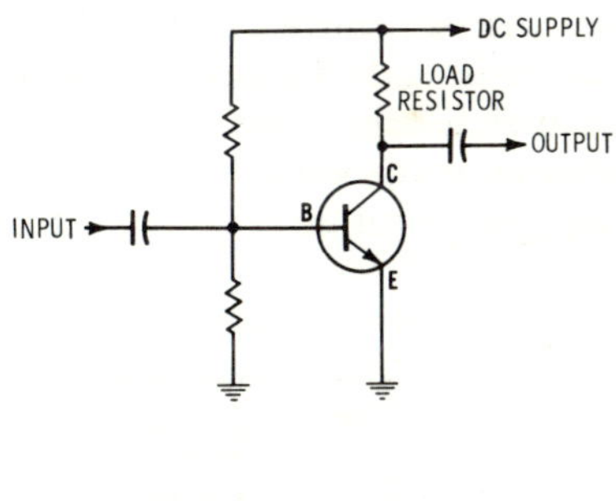

Fig. 1-10. Voltage-divider bias.

Voltage-Divider Bias

Fig. 1-10 shows the most popular circuit, one probably used more than any other. This uses a standard voltage divider with two resistors from the dc supply to ground. The transistor base is connected to their junction. By changing the ratio between the two resistors, we can put any voltage we want on the base of the transistor. The voltage-divider action will "clamp" the base at whatever voltage is desired.

These resistors are made large enough to hold the input impedance of the stage to whatever level is needed. The steady-state bleeder-current that flows at all times in the divider will be set at a value that is considerably greater than the comparatively small base-emitter current. So, the divider will stabilize the base voltage. Base current will be only a few microamperes, and the bleeder current may be several milliamperes.

The Emitter Resistor

In Fig. 1-11, we've added a resistor that a lot of old tube-type technicians would call a "cathode resistor." It is, in a way, but this is not its main purpose. This resistor is used mainly to *stabilize* the transistor. All transistors are highly sensitive to temperature; if the junction temperature rises a few degrees, the current rises with it. This can change the gain of the stage and, worst of all, it can start a sort of chain-reaction, in which the transistor keeps on getting hotter and hotter and drawing more and more current until it destroys itself. This is "thermal runaway." So, we add this little resistor to prevent it.

As the collector current rises (which it will do if the junction-temperature rises) there will be a greater voltage drop across the emitter resistor. Since the transistor bias is between emitter and base, this will act to reverse bias the transistor, and thus cut down on the excessive current. While this resistor does have an effect on the bias, it is primarily a safety feature.

Now, of course, the actual bias on the transistor is the voltage between emitter and base—not the voltage from base to ground. Let's say that we started with a divider voltage of +1.0 volt applied to the base, with the emitter grounded. If we add an emitter resistor of such a value that it develops +1.0 volt, we have both base and emitter at the same voltage, and the transistor goes to zero bias and cuts off. What do we do to get rid of this? No problem! We simply

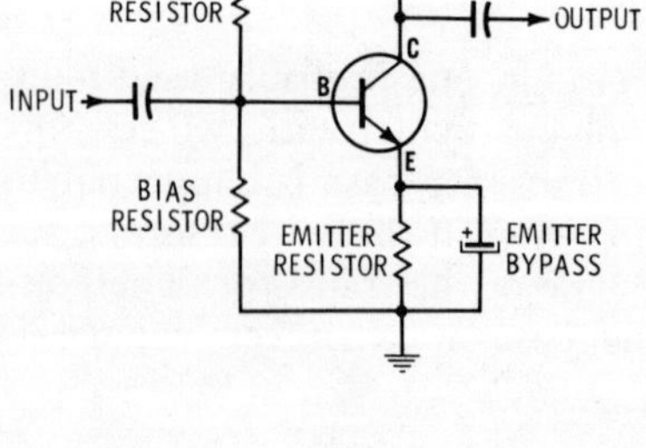

Fig. 1-11. Bias circuit with emitter resistor added.

make the base voltage a little more positive, so that the base is now +2.0 volts to ground, and we're back where we want to be; +2.0 volts minus +1.0 volt equals a net base-to-emitter voltage of +1.0 volt.

This emitter resistor is going to be very important to us as we go along. So remember it and how it reacts. This one little resistor is going to be one of our most helpful tests in locating all kinds of trouble.

DIRECT-COUPLED CIRCUITRY

Now, we're going to get into a new one. In most transistor circuits, you can find a "tube analogy," exactly the same circuit but with transistors instead of tubes. However, you're going to run into one very popular circuit that hasn't had a tube analogy since about 1928. (There's nothing really *new* in this world, just new uses for old ideas!) This is the "direct-coupled" circuit, where "everything is tied to everything else." These are the hardest to understand, and can be the hardest to service unless you know exactly how they work and what's going on inside of them.

Therefore, let's take a close look at the basic circuit and see how it works. Once you understand the two-stage circuit, you have got them all. In more complex circuits, they just add more stages, that's all. You will find some amplifiers with as many as five or six stages, all directly coupled, but they are all alike after the first two stages.

Fig. 1-12 shows a two-stage preamplifier-driver circuit as used in a General Electric T-50 amplifier. The dc voltages shown on the transistor elements are the no-signal voltages. Q1 is an npn silicon transistor, connected as a common emitter, and biased in Class-A.

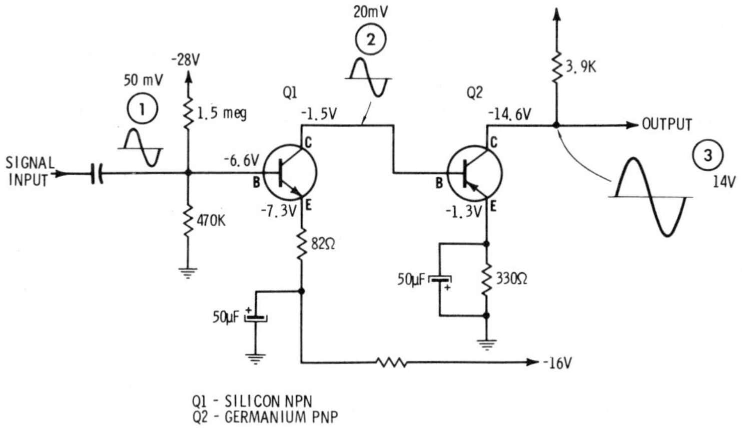

Fig. 1-12. Direct-coupled preamplifier and driver.

Its collector is directly connected to the base of Q2, which is a germanium pnp transistor, also biased to work Class-A. Note the bias voltages shown. Q1 has −7.3 volts on the emitter and −6.6 volts on the base. (The emitter voltage comes from the dc supply through a dropping resistor.) Subtracting, this gives us $7.3 - 6.6 = 0.7$ volt, negative. Silicon transistors have a minimum working bias of 0.6 volt, so this gives us just enough extra forward bias to make the transistor work in Class-A, higher on the curve.

On Q2, we see −1.3 volt emitter and −1.5 volt base; this gives us a net of −0.2 volt, which is the normal bias for a germanium transistor. With the application of a signal, these voltages will rise slightly, putting both transistors at the point of good Class-A operation. Now, what happens, otherwise, when a signal is applied?

If we feed in a standard test signal at the 50-millivolt level (50 mV) to the base of Q1, it will be amplified and appear at the collector, inverted in polarity. However, the load resistor of Q1 now isn't an actual resistor, but the base-emitter impedance of transistor Q2. This is a pretty low impedance, as are all common-emitter circuits. So, we will not find a *voltage* gain across Q1. In fact, we'll lose a good bit of voltage, as you can see from the 20-mV signal shown on this connection.

However, there will be a *power* gain; this *is* a current amplifier. This will cause an increase in the base-emitter current of Q2, which in turn causes a greater current in the collector-emitter junction, and we have amplification.

Q2 works into a good-sized collector load resistor, of 3900 ohms, just as in previous circuits. So, now we go from current amplification and back to voltage amplification. The signal now rises to 14 volts peak-to-peak. Actual gains in this circuit are about 30 (power-gain) for Q1, and almost 250 (voltage) for Q2. In the complete circuit, the collector of Q2 is connected directly to the output transistors. These circuits will be dealt with in the following chapter.

The advantages of direct coupling are a reduction in the number of parts needed and the complete elimination of all "signal coupling devices"—transformers, capacitors, etc., from the signal path. The frequency limitations that would be imposed by these components will be absent. Direct-coupled amplifiers can be made with a tremendous frequency range, from very high frequencies down to dc in many instances.

Direct-coupled stages can be tricky to check out because of the interaction between stages. As you can see, everything affects everything else. A fault in the first stage, such as a transistor with a little too much leakage, upsets every stage from that point on; even dc voltages are affected. In the chapter on testing, we'll give you several simple tests that will make life easier.

The big disadvantage of direct-coupled circuits, as far as the service technician is concerned, is that he cannot use an in-circuit transistor tester with them with reliability. If you'll look at Fig. 1-12, you'll see why. If you make your connections to Q2, for example, you will also have the emitter-collector junction of Q1 connected directly across the base-emitter junction of Q2. A fault-reading here could mean that either one was bad, or, in a slightly different circuit, that one of the junctions was forward-biased and thus reading as a short or leakage on the instrument.

Dc voltage readings are a very useful test for these circuits. Any fault will upset the dc voltages. Go back from the output toward the input; find the point where the dc voltages are normal again, and start there. As we have said, any fault in an early stage affects every stage from that point on toward the output. The circuit shown in Fig. 1-12 is one channel of a stereo amplifier; the other one is exactly like it. Since identical faults seldom (but not *never!*) happen in both channels of such an amplifier at the same time, you can often use the working channel as a working model in checking out the dead one. Compare dc voltages, signal levels, etc., between the two, and you'll be able to locate the area of the trouble much sooner.

TONE CONTROLS

Tone control circuits used will be just like those found in tube amplifiers. In the circuit shown in Fig. 1-13 as used in a Magnavox R-212 amplifier, the bass control is a 3.3 megohm variable resistor. It is connected just "behind" the .02μF coupling capacitor from the audio input. This .02 μF capacitor has a reactance of about 150,000 ohms at 100 Hz. The low frequencies divide across the coupling capacitor, the bass control, and the 33k resistor to ground. If the bass control is set for maximum resistance, the greater portion of the lows will be developed across this control and fed on into the amplifier. With the bass control at minimum resistance, most of the low frequencies will be dropped across the capacitor. The 33k resistor is used to prevent the bass control from causing too great a drop in signal amplitude, when set at its lowest resistance value.

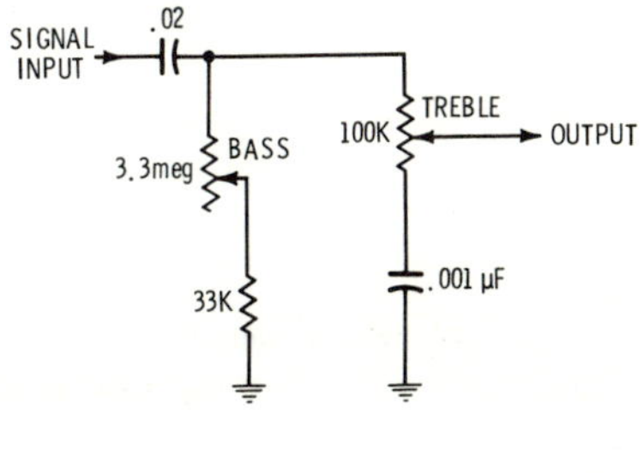

Fig. 1-13. Bass and treble tone controls.

The high-frequency response is controlled by a simple RC network, made up of the treble control and its .001 μF series capacitor. This is the same old "high-cut" circuit we've used for years. The smaller .001 μF capacitor has a high reactance at low and midfrequencies, and a much lower reactance at high frequencies. If the slider of the treble control is moved to the bottom end, nearest the capacitor, most of the highs will be shunted through this capacitor to ground. Putting the treble control at the top end allows the highs to go on through, since we now have maximum resistance in series with the capacitor.

THE LOUDNESS CONTROL

The standard volume control is simply a high-resistance signal-voltage divider, placed between the output of a signal source (microphone, preamplifier, etc.) and a following amplifier stage. Fig. 1-14A shows the standard circuit. You'll find this in all of the simpler amplifiers. However, when we get into high-fidelity amplifiers, this circuit has one defect. When the slider is turned to the lower end of the resistance, it has a tendency to cut high frequencies, due to the reduced resistance to ground.

As a result, a "frequency-compensated" volume control circuit was developed for such critical applications. This is shown in Fig. 1-14B. This is called a "loudness control" to distinguish it from a plain old volume control. The RC networks on the resistance taps help to maintain the correct frequency response, balanced between highs and lows, no matter where the control slider is set. The capacitors feed more highs to the lower end of the resistance, eliminating the low-cut effect.

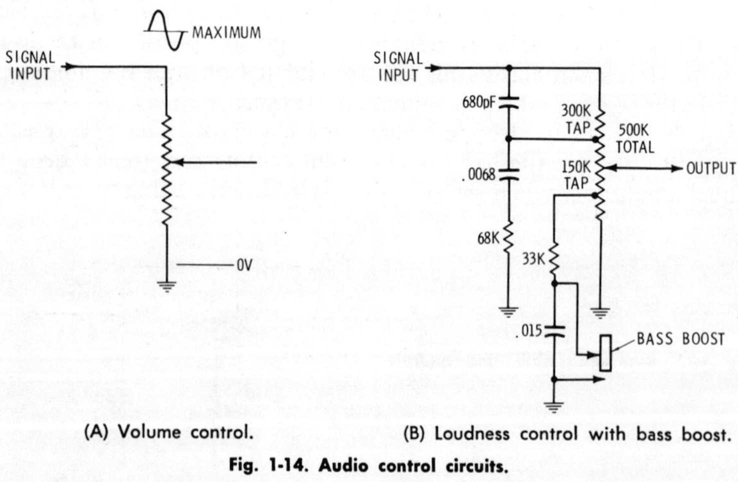

(A) Volume control. (B) Loudness control with bass boost.

Fig. 1-14. Audio control circuits.

You may find extra devices on these circuits. Note the slide-switch shown connected to the .015 μF capacitor on the lower tap, marked "bass boost." This is used to raise the proportion of lows, near the lower end of the control, for low-level listening. Again, this is the same high-cut circuit we just saw used as a treble control.

MONO VERSUS STEREO AMPLIFIERS

Up to this point, we have been talking about single-channel amplifier circuits. These are now, somewhat incorrectly, called "mono," short for "monaural." (That is a misnomer. You do *not* "hear with one ear," which is the literal meaning of the word "monaural.") Anyhow, it's a nice short word, and we have to call the circuit something. The term is used to distinguish between single channel amplifiers and stereo, which are dual-channel amplifiers. ("Stereo" isn't too good a word, either, but everyone uses it, so we'll go along with the crowd.)

You'll find stereo amplifiers everywhere now—radios, phonograph amplifiers (which have converted almost universally to stereo, especially in the larger types), and even in some musical-instrument amplifiers that have two separate channels and thus fit the definition of stereo. This makes things very handy for the service technician. With certain exceptions, which will be duly noted and explained in a later chapter, any trouble in one channel of a stereo amplifier can be quickly checked out and pinned down by using the other channel for direct A-B comparisons; dc voltages, signals, gain-per-stage tests, and so on. Any trouble that affects both channels at the same time is most apt to be in the one section that is common to both, the dc power supply.

Each stereo amplifier consists of two identical channels. When we discuss circuits, up to this point and in later chapters, we will work with only one channel. The other one will be exactly like it; only the part numbers will be different. Each channel of the larger units has its own individual loudness control, bass and treble controls, etc. In a few you'll find dual controls for loudness, but they will usually have a split knob so that the gain of each channel can be controlled separately.

The only different control you'll find in most stereo systems will be called a "Balance" control. Fig. 1-15 shows how this is connected in a General Electric T-50 stereo amplifier. There may be dual controls in some of the more elaborate systems, but they all do the same thing, "balance" the gain between the two channels to compensate for any possible unbalance in the recording, tape, or whatever signal is being used. It will also let the listener set the system up to fit his particular preference in reproduction.

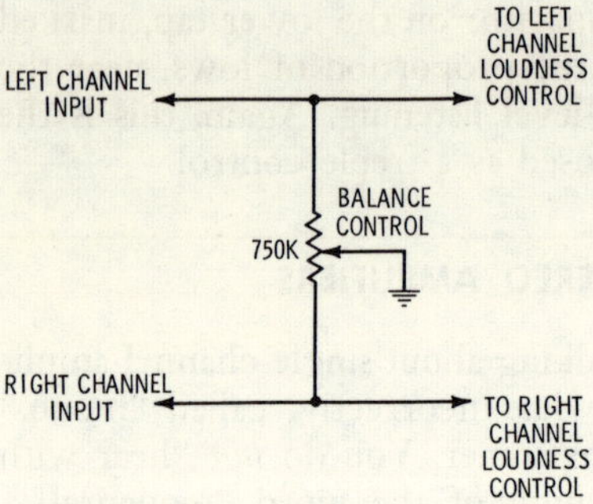

Fig. 1-15. Balance control in stereo amplifier input.

When this balance control is correctly set, the stereo effect or "presence" will be much better. You can check the operation of the system and this control with a special test record. This has a sharp "tick" sound, like a metronome, which is recorded with exactly the same amplitude in both channels. While playing this band on the test record, stand midway between the two speaker systems, close your eyes, and adjust the balance control until the tick sound appears to come from a point exactly *between* the two speakers. This effect means that each channel has exactly the same gain.

Electrically, the operation is simple, as you can see. If the balance control is set to its exact center, and each channel has precisely the same input signal level, the output of both will be the same. If the listener wants to make one channel slightly louder than the other, he moves the balance control *away* from that channel input. This gives this channel higher resistance to ground, raising its signal level, while at the same time putting a lower resistance to ground in the other, reducing its signal level. Most balance controls will cut off one channel or the other completely, when the control is turned all the way to one end.

This is a very useful quick check for the technician. By playing a mono record (equal signal in both channels), or listening to a mono radio program, etc, he can turn the balance control all the way from one end to the other, and thus get an idea of how they are both working. This will show up weakness in one channel or even a "one-channel-dead" problem. There are several test records that have a tone signal recorded in one channel only. This is very useful for checking stereo separation, which means the ability of the stereo amplifier to reproduce a signal in one channel while keeping it completely out of the other. Normally, the separation is seldom perfect; there will almost always be some feedthrough between channels, due to the circuitry. The better amplifiers have very good separation, while the cheaper ones are not nearly so good in that respect.

CHAPTER 2

Drivers and Output Stages

Up until now, we've been talking about voltage-amplifier stages, preamplifiers, etc. This is what we'd call them in tube circuits. In transistor circuits, they're just called "small-signal" circuits, for obvious reasons. You'll find low voltages and pretty small currents, and low-power rated transistors.

Now, we're getting to the output stages, and power amplification. In some types, such as auto radios, the voltage will still be pretty low, but the currents will be much greater; so, the total power (E × I) will be much larger. In fact, the old tube-type technician will find currents that will amaze him. Some of the bigger amplifiers use currents up to 10 or 15 amperes, whereas he's used to working with milliamperes. This brings up a few special problems, which will be dealt with later.

DRIVER STAGES

The driver stages are included here, because they are about halfway between the small-signal types in earlier stages and the power output stages. You might call these medium-power stages. Voltages will be pretty low, but the current will be higher. These stages take the signals after they come from the preamplifiers, and raise them to the level needed to drive the output stage to full rated output.

Input Power

There's one big difference between tube and transistor power amplifiers. This is in the amount of *power* needed to drive the output stage. In a tube, no matter how great the power output, there will seldom be any power used in the grid or input circuit. (High voltage, but hardly any current, except in very special cases.) Here, a voltage on the grid controls a large current in the plate circuit. In transistor circuits, the base current controls the collector current. (Don't worry; we're not going back to that "current-amplifier" bit. We are, and always will be, able to use the *voltage* as an indicator of conditions in any circuit. A given current in the base circuit will produce a given voltage drop. If this voltage drop is right, the current is right, and that's that. That is, unless one of the resistors is burned up, and that's pretty simple to check out.)

Transistors used in driver stages will be slightly larger, or be able to handle a little more current than the ones we've been discussing before. Ball park figures for a driver stage will be about 2 watts. Transistors used will be rated at about 5 watts maximum. These things must be taken into account when you're selecting replacement transistors; otherwise, there will be no difference.

Coupling

You'll find three types of coupling circuits used. Two of them will be familiar (see Fig. 2-1). At Fig. 2-1A is the old-faithful, transformer-coupled in and out stage. The driver transformer is used to match the higher impedance of the driver output circuit to the comparatively low impedance of the output-stage base, for most efficient power transfer. Normally, we'd have to match impedances in tube circuits only between the output-stage plate and the speaker, since this is the only point where we're dealing with any power. However, here we have a certain amount of input power too, so it is necessary to match again.

At Fig. 2-1B is another familiar circuit with capacity coupling. The bias of the output transistor is set by the voltage divider resistors from the collector, and for dc the capacitor blocks this from the driver stage. The main difference from tube circuits will be by comparison in the very large values of capacitors used. Where tube circuits use only fractional units, say up to .02 or even 0.1 μF, transistor circuitry will use 4, 8, or 10 μF and they'll be low-voltage electrolytic capacitors. (These can cause some "funnies" as we'll see later, so don't forget them.)

At Fig. 2-1C is a circuit that looks new, but isn't. It was used as far back as 1928, in one or two makes of radios. This is "direct coupling," and it means just what it says. The collector of the driver transistor is tied directly to the base of the output transistor. So, the driver collector current is also the base current of the output stage. Any "signal variations" in the driver output will be transferred directly to the output stage base. This is somewhat more efficient than previous circuits, and it certainly does save on parts.

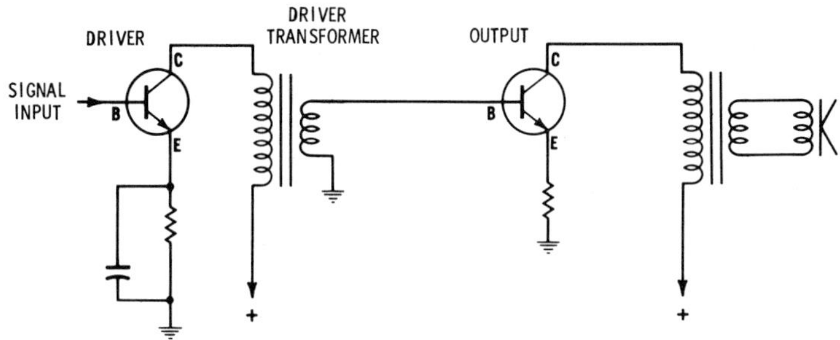

(A) Transformer coupling.

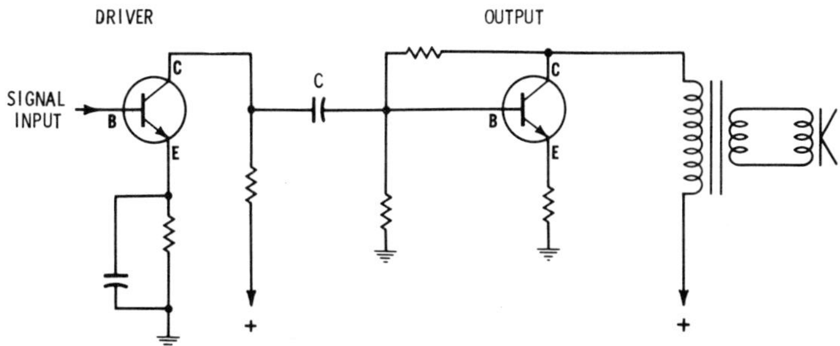

(B) Capacity coupling.

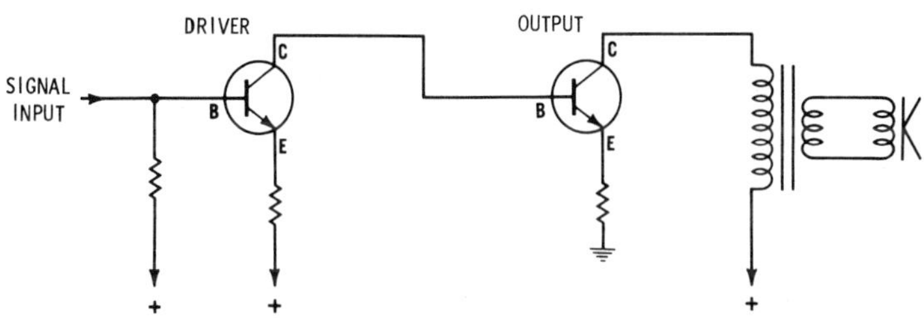

(C) Direct coupling.

Fig. 2-1. Interstage coupling circuits.

However, there are a few peculiarities in this type of circuit that you will find. Since the collector current of the driver transistor determines the base bias of the output transistor, any leakage or defect in the driver can cause a much greater change in the output transistor characteristics. This can lead you into thinking that there is still trouble in the output stage even after you have replaced the output transistor.

A shift in characteristics, or a slight mismatch between a new output transistor and the original driver transistor, can upset bias and lead you astray down the wrong path. If you run into problems after output transistor replacement in this circuit, pull the driver transistor and check it or try a new one. The voltage developed as bias for the output transistor is directly related to the current-gain of the driver transistor, its "beta." If this is too high or too low for the new output transistor, the bias will be upset. Try a lower beta driver and see if the bias goes back to normal. Later on we'll give you some tests to pin this down.

In some of the small amplifiers, you may find that the whole thing it direct coupled, four or five transistors, all tied directly together from input to output. These can be bothersome to check out with conventional methods. Any small problem in an early stage will cause bias upsets and troubles all the way through the whole circuit; everything "after" this point will be wrong. There are two ways of checking these. One is to go back through the circuit until you find a place where the voltages are normal, and look there. The other is to signal-trace through the circuit until you find where the signal stops, and check from there on. These cannot be checked reliably with an in-circuit transistor tester. You always have at least one junction of another transistor shunted across the ones you want to check. The simplest way to do this is to take one out somewhere near the middle of the circuit, thus "breaking the chain"; now, you may be able to check a few more. Definite proof will usually be found only by taking each transistor out and checking it. Dc resistance readings in these circuits are a waste of time. There are so many shunt paths and transistor junctions that you can't get any useful information with this method. Disconnection of one end is the only way of checking any resistor used.

SINGLE-ENDED OUTPUT CIRCUITS

When you get to checking around the output stage, one of the things you need to know is what class of operation it is designed for—that is, whether it is supposed to be working Class-A, Class-B, and so on. This will tell you what to expect in bias voltages, waveforms, and other circuit behavior.

Class-A Amplifiers

All of the previously discussed stages have been working in Class-A. This type of amplifier has two distingusihing characteristics. One, and most important, the waveshape of the output signal must always be *exactly* the same as that of the input signal. If it isn't, you've got some bad distortion. So, this determines the operating bias; the transistor must be biased up on the curve to the point where the signal can swing above and below this level, reproducing the input signal exactly (Fig. 2-2).

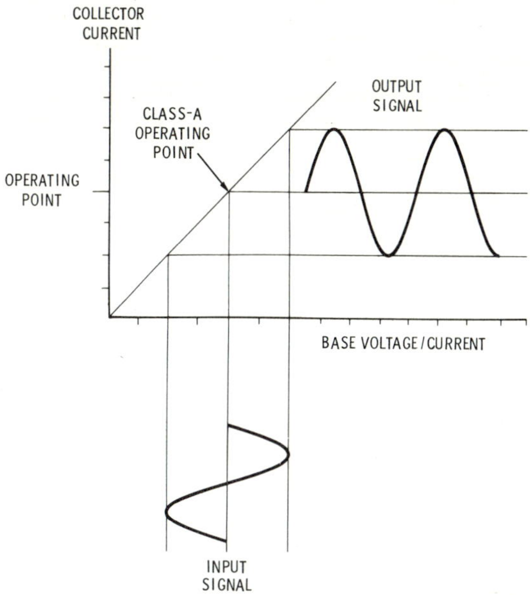

Fig. 2-2. Class-A operation.

This is nice, but not particularly efficient. The main problem in its design, is the need for a certain amount of collector current at all times, even when the stage isn't amplifying a signal. So, this "resting current" is actually wasted, unless the stage is working. In some stages, this could be considerable. However, for reasons of economy, space, etc., you'll find quite a few single-ended Class-A stages used; these are mainly found in older auto radios and very small phono amplifiers. Remember this fact: If there is only *one* output transistor used, this must be a Class-A stage.

Fig. 2-3 shows a typical single-transistor, transformer-coupled, output stage, something like those found in auto radios and small amplifiers. This is actually a common emitter circuit, but notice the voltage-supply point. The collector is grounded, through the primary winding of the output transformer, and the voltage is fed to the emitter. This is exactly the same as the other common-emitter circuit, with voltage fed to the collector and the emitter grounded. If you read the voltages as you should, from the transistor emitter, you will come out with the same voltage and polarity as before.

Both types are common-emitter. The grounded-collector circuitry is used to keep any dc voltage off the speaker voice-coil wiring, for safety in auto-radio uses, etc. You can always tell a common-emitter circuit from the other grounded-collector circuit, the emitter-fol-

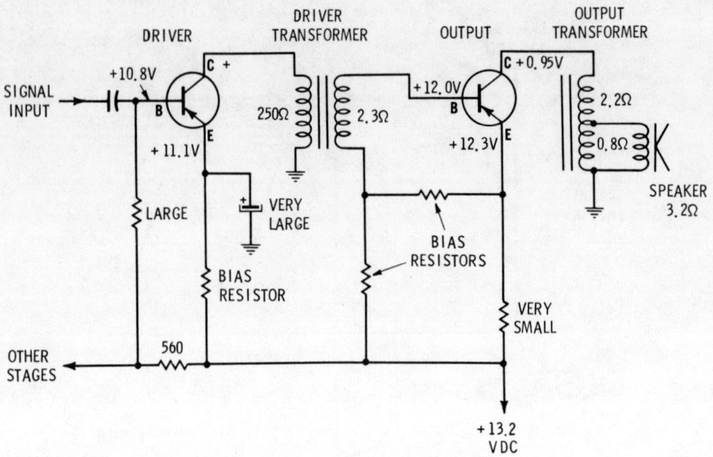

Fig. 2-3. Single-transistor Class-A output stage as commonly found in auto radios and other amplifiers.

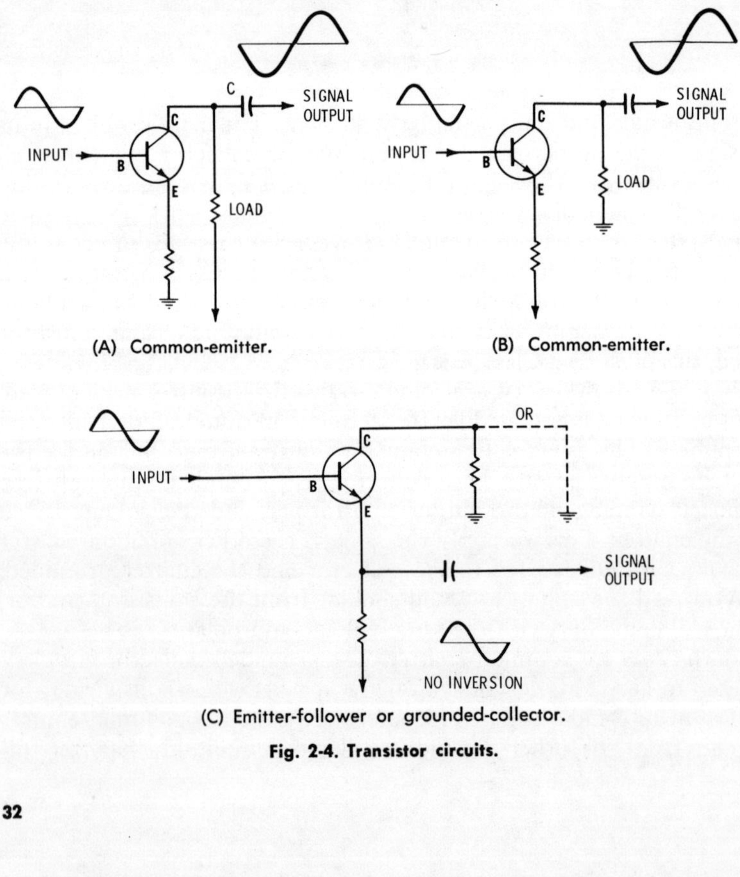

Fig. 2-4. Transistor circuits.

lower, by looking to see where the *signal* is taken off. Fig. 2-4 shows the difference. If the signal is taken off at the collector, this is a common-emitter circuit, no matter what it looks like. If the signal comes off the emitter, it's an emitter-follower.

The output stage bias is set by a network of resistors from the base and emitter, going back to the voltage supply. The emitter resistance, a very small one, usually less than one ohm, is used for the same purpose as before. To prevent the transistor from going into thermal runaway, it counteracts any changes in collector current due to junction heating.

The dc voltages as shown in Fig. 2-3 are no-signal voltages, taken from an actual circuit (Mopar 109 auto radio). With a signal applied, they'll change slightly. Notice the difference in dc resistance of the driver transformer primary and secondary windings. This shows the large step-down in impedance needed between these two circuits, to match the higher impedance of the driver collector circuit to the much lower impedance of the output transistor base for most efficient power transfer.

The actual resting current of this stage will be somewhere around 1.0 ampere. It won't vary a great deal with an applied signal, since the addition will be in the form of ac, and the average current will not show on a dc ammeter. Remember, the output transistor is biased on its collector-current base-voltage curve to the point where it draws current at all times.

Note the bias on the base-emitter; this is about −0.3 volt. Since this is a pnp transistor, the base must have a certain amount of forward bias to let it draw any current at all. So, about −0.3 volt is applied by the resistors. This tells us something else; that this is a germanium transistor. Germaniums need a minimum of 0.2 volt to put them into conduction; silicons require 0.6 volt. To make the transistor draw a little resting current and work in Class-A, an extra 0.1 volt of bias is used. To the old tube-type technicians, a tenth of a volt won't sound like very much bias. But in transistors it is. It takes only a few tenths of a volt bias-change to cause a very large collector-current change in transistors.

The output transformer in this circuit is actually an autoformer, with the speaker voice coil being tapped down on the winding to the point where it matches. However, there is no difference at all between this circuit and the two-winding or "full transformer," either in characteristics or in output. The autotransformer is simpler and cheaper.

This type of circuit has one characteristic that you *must* look out for when servicing it. NEVER turn one of these radios on without a speaker or dummy load connected. If you do, the alteration of the load impedance will cause the voltage across the output transistor to go way up, and breakdown of the collector-emitter may possibly destroy it.

Transformer-coupled transistor output circuits will always tolerate a short across the speaker terminals, but they do not like open circuits. In many auto radios, you'll find a special shorting-type speaker jack; when the speaker is disconnected, the voice-coil terminals of the output transformer are deliberately shorted. This is a safety feaure.

For bench servicing, you must either use a duplicate of the original speaker plug or break the short in the jack by slipping a piece of paper between the contacts. For testing, speakers with exactly the same impedance as the original must be used. These will vary from 3 to 4 ohms up to as high as 45 ohms or more; check the schematic to make sure of the impedance of the one you have. (If you think this is odd, wait until you see some of the later circuits.)

Test Characteristics

This type of circuit has some very useful characteristics for testing. Notice the very small *voltage* on the collectors of both driver and output transistors. Even though they go directly to ground through their transformer windings, there is a very minute dc voltage developed *if* the transistor is conducting. This, as you'll see time and again as we go along, is the one thing you must find out in all transistor circuits: Is this transistor *conducting current?* When you do find a bad transistor, or one that isn't working for other reasons, it won't be conducting; it will be either cut off, open, or shorted. So, what we want to know is, is the thing conducting?

Here, there is no voltage applied to the collectors. If there is any voltage at all present, this means that it is a *dc voltage drop* across the very low resistance of the windings. This is normal. As a typical example, if the radio is completely dead–no sound at all–and you read 1.1 volt on the collector of the output transistor, and zero voltage on the collector of the driver, where is the trouble? Right! The driver transistor is not conducting. The output transistor is probably working very well. You can quickly verify this conclusion by feeding a test signal into the base of the output transistor, by pulling the driver transistor and checking it, or even by breaking the circuit and measuring the current through the driver transformer primary. Frankly, the last step is seldom needed.

Problems that could cause such a condition might be: an open driver transistor; open emitter resistor; no dc voltage supply to base-emitter; open primary winding on driver transformer (although if you're using a vtvm or tvm this would be more apt to give you a fairly high dc voltage reading on the collector); incorrect base-bias due to resistor drift; or a break in the PC board somewhere.

These are all comparatively easy to check. First, check the dc voltages on the base and emitter. Since these are usually set or clamped by resistance voltage dividers, they should be pretty close to normal if the transistor is open. If so, this clears up the chances of a broken board, open resistor, or that type of trouble. Dc voltage much below normal on the emitter could indicate an emitter-collector short in the transistor. If the short were bad enough, this would probably give you a collector voltage quite a bit higher than normal.

Check the bias carefully. Watch out for cases where you find exactly the same voltage on base and emitter. This always indicates *zero bias* on the transistor, which means that it is completely cut off. The only time you'll ever find a cut off transistor in normal use is in tv circuits (sync clippers, etc.). As far as I know, you will never find this in audio work. The bias may be only 0.2 volt, but it had better be there. Incidentally, if your voltmeter will not read tenths of a volt accurately, don't despair. Just open the emitter or collector circuit, and connect the vom in series as a milliammeter or ammeter. The voltage may be too small to read accurately, but the *current* won't be, especially in audio output stages. If there *is* current, the stage is working or trying to work. If you don't know about what the current should be, start on your highest current range, then switch down until you get the reading near the middle of the scale. Some of these things draw pretty big currents.

PUSH-PULL CIRCUITS

Now, we come to a circuit that is probably the second oldest in audio, the push-pull Class-A, which uses two active devices (tubes or transistors, makes no difference) connected to either end of the output transformer primary. Fig. 2-5 shows a typical circuit used in a pa amplifier.

Class-A Amplifiers

Let's run over the important characteristics of this circuit. You'll not only have to know them for use in servicing, but you'll need them as a basis for working on other similar, but far from identical, circuits that you'll see very soon.

First, there are two transistors, one connected to each end of the primary winding. Both of these work Class-A. That is, they are drawing current at all times during one full cycle of the signal. Signal currents in their collector circuits are exact replicas of the input-signal waveform. However, the two transistors work 180° out-of-phase. As the current in one rises, the current of the other falls. Note the output waveforms.

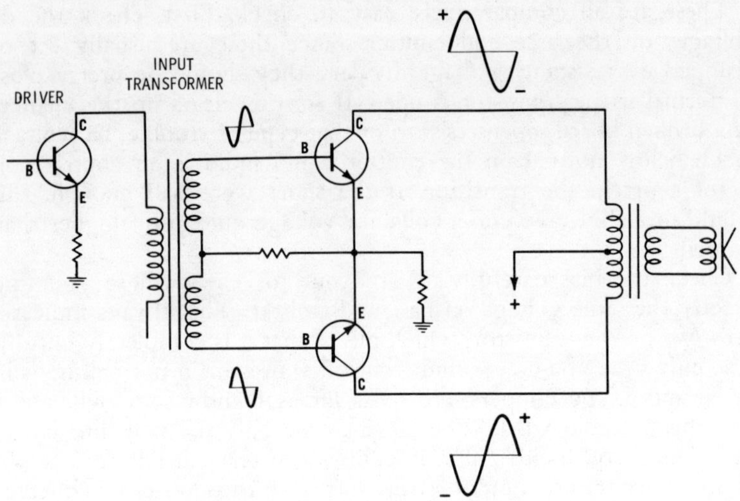

Fig. 2-5. Basic Class-A push-pull circuit.

If you could say that a positive-going half-cycle is "pushing" current down through the primary from the top transistor, then the lower transistor has a negative-going half-cycle and is "pulling" current out of the primary. This is where the name comes from. This circuit is more than twice as efficient as two transistors of the same type connected in parallel and has other advantages as well.

One major requirement of a push-pull, Class-A circuit is two out-of-phase *input* signals, as you can see. Since the output transistors are common-emitter circuits, they invert the signal. So, the base signals must be out-of-phase to provide the necessary drive. Here, this is done by taking the base signals from opposite ends of the secondary winding of the driver transformer. (Opposite ends of a transformer winding are always 180° out of phase.) This is the simplest way. Later on, you'll see several other ways of getting this phase-inversion. No matter how we get them, they all do the same thing.

Since both transistors in this circuit are conducting at all times during the cycle, one of them can fail—open, for example—without completely killing the input signal. If one of them quits, the other one goes on pushing and pulling, and the only result is a loss of volume. Remember this reaction; you're going to find other circuits with a completely different fault-reaction, later on.

In Fig. 2-5, the collector current is fed to the center tap of the output transformer primary. Speaker loading in this kind of circuit is almost as critical as in the single-ended stages. Like the single-ended circuits, this one should never be turned on unless there is a speaker or a dummy load connected to the output. If it is turned on

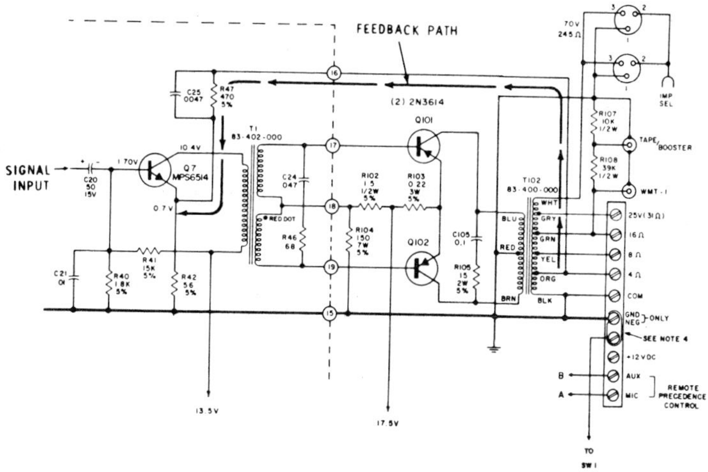

Fig. 2-6. Output stage of Bogen CHS20 amplifier.

unloaded, the voltages across the output transistors may rise to such levels that the collector-emitter voltage-breakdown rating is exceeded, and BANG. (Average elapsed time, about 5/100 of a second, or maybe less.) So, make sure that the speakers are tightly connected before you apply the power.

Fig. 2-6 shows the complete circuit of the original, as found in a Bogen Challenger CHS20 public-address amplifier. As is customary in this type of amplifier, several impedance matching taps are provided on the output transformer secondary, so that any desired combination of speakers can be hooked up and properly matched.

Here again, we see the grounded-collector, hot-emitter type of power supply to the output stage. As we have said and will say again, this makes no difference in operation at all. As long as the transistors "see" the proper voltage and polarity, fine.

The Feedback Loop

One of the major advantages of the "double-ended" type of output circuit is its cancellation of second-harmonic distortion. This is due to the action in the transformer primary. For a little more improvement in frequency response, and to hold down odd-harmonic distortion (third, fifth, etc.) degenerative feedback is also used. Note the connection to the 4-ohm secondary tap of the output transformer. A small signal is taken from this tap and fed back through an RC network to the emitter of the driver transistor.

These low-level signals arrive at that point in "antiphase," and thus are degenerative (tending to stop oscillation). This degenerative action also helps to cancel out the distortion. Because the feedback signals are out of phase with the input signals, they also cause a small loss of gain. However, the improvement in the overall frequency response of the amplifier is well worth this sacrifice. In fact, the loss is normally made up by designing the original circuit for a larger "open-loop gain" (without negative feedback) than is necessary. When the feedback is applied, the amplifier will still come out with the rated amount of gain and power output. Gain can easily be made up, but frequency response is harder to get.

As we just said, the feedback signals are out of phase. What would happen if they got there in-phase or regenerative (tending to cause oscillation)? That's right, the amplifier would oscillate. Normally, this cannot happen unless the output transformer has been replaced, and the primary connections have been *reversed.* If the amplifier breaks into oscillation as soon as it's turned on after output transformer replacement, turn it off (quickly!) and reverse the collector leads from the transformer. This will invert the phase of the feedback signals, and restore the amplifier to normal operation.

OUTPUT TRANSFORMER REPLACEMENTS

Output transformers in tube amplifiers are easy to replace. You look up the load impedance of the tube types used, the speaker voice-coil impedance, and the power rating, and there you are. There are hundreds of transformers that will work perfectly. However, there is a great variation in transistor output (and driver) transformers. As far as we have been able to find out, there is no such thing as a "universal" transistor output or driver transformer; that is, one which can be adapted to match a good many transistor amplifiers, by changing taps, etc. Each one is specifically designed to work with the transistors used in that circuit, signal-voltage levels, impedances to be matched, and so on.

So, if you should find a defective output or driver transformer, your best bet would probably be to order an exact duplicate replacement transformer from the original manufacturer of the equipment. If the amplifier was built in the U.S. by a reputable maker, this will be easy. In fact, some of the replacement-transformer manufacturers, who list thousands of transformers in their catalogues, may even have duplicates for the U.S. types; check by make and part number.

However, if the amplifier is an import, you may be in deep trouble. You might have to do a lot of very careful cut-and-try work with different transformers before you locate a suitable replacement. No guarantees that you will ever be able to find one, of course! Incidentally, if this unpleasantness should happen to you, there is one method that will be quite helpful. After you have completed "figuring it out" and chosen the best possible replacement, try the amplifier out with current meters installed in vital places, and at less than half the normal input voltage! This can be done with a Variac or step-down transformer. The reduced transistor voltages will allow you to monitor the currents, which will be kept down below dangerous levels. You can start at almost zero volts input, and gradually raise the power, watching the currents, and make any corrections or adjustments needed. Transistors can blow so fast that something like this can be a very useful service technique. In fact, you can use it after any kind of repair, such as output transistor replacement, where you'd like to make sure that your new $15.00 replacement transistor wasn't going to go Pow as soon as you turn it on!

The replacement transformer makers, even now, list a good many transistor transformers, in all sizes and power ratings. It is quite possible that they'll come up with open stock replacement parts for the better-grade import amplifiers. I certainly hope so, for your sake and mine too. Check their catalogues, by maker's name or brand name—not necessarily the same—and part number. However, from information available at this writing, you will not find replacement transistor transformers as simply as you can locate tube types. No more saying "Gimme a 10-watt output transformer for two 6L6s to a 16-ohm voice coil!"

There is one ray of sunshine, though. Because of this situation, but mainly for the sake of economy, very few transistor amplifiers use output or driver transformers at all. In fact, you'll find that the great majority of transistor amplifiers now, even the very small and the very large, use a special circuit called "Output-Transformerless" or "OTL" for short. This uses two transistors, but they couple directly to the speaker.

Ray-of-sunshine No. 2: Tube output transformers are very high impedance, wound with very fine wire, and carry comparatively low currents. They are subject to damage from electrolysis if the enamel on the wire is broken, and to burnout from overheating if a tube shorts, and so on. On the other hand, practically all transistor transformers are wound with comparatively large wire because they have to carry high currents. Their lower impedance requires fewer turns of wire, too.

If there is a short in the circuit that draws an overload of current through the transformer, the chances are that it will stand up until the fuse can blow or the circuit breaker can pop open. Actually, I can remember only one case of output transformer trouble, and this was in an auto radio quite a while back. The tap on the autotransformer output was open. This was found to be due to a poor solder joint where a loop of wire had been pulled out of the winding, and clipped, to make the tap. Resoldering the two wires together fixed that in short order. So, if you do find an open transformer winding, look for something as simple as this. You just might find it.

Reverse

For the last example in the transformer-coupled stages, look at Fig. 2-7. Here is exactly the *same* output stage we have been working with, but check the connections. The driver transformer connections are the same, and the dc voltages on the transistors are the same (reversed polarity, since this one uses npn transistors). But, look at the collectors of the output transistors. They're tied together and have the high voltage, +35 volts, applied directly to them. How in the world do we get a signal out of this?

Look at the emitters. One is tied to each end of the output-transformer primary winding, and the center tap of this winding is grounded. Since the current here is actually the "emitter-to-collector" current, we have exactly the same current through the primary winding, and the operation of the stage is identical to those

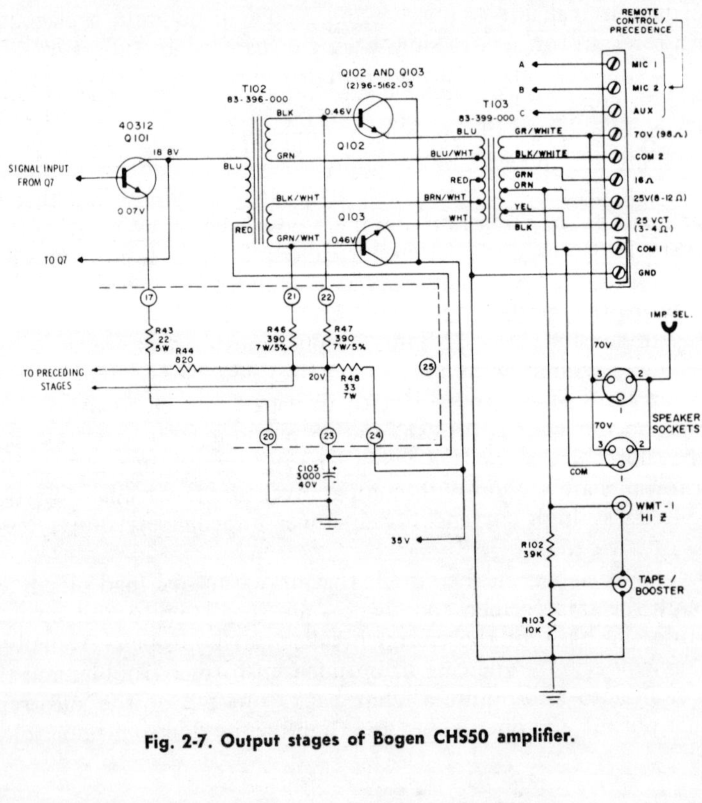

Fig. 2-7. Output stages of Bogen CHS50 amplifier.

discussed previously. The reason we brought this out specifically is that a great many persons get confused by the reversed connections (apparently!) that come up in transistor circuitry. However, if you will simply follow the collector current, from the power supply through the emitter, to the collector, and the load, then back to the other side of the power supply, you'll see how it works. It makes no difference *where* the load (output transformer primary) is in the circuit, as long as the signal currents flow through it. That's all we want!

CLASS-B OPERATION

Let's make our Class-A stage more efficient. Say we have a pair of transistors, each capable of a 5.0 ampere collector-current swing. In the Class-A stage, we had to bias them to a point in the middle. So, each transistor has only a ±2.5 ampere swing. This is 5.0 amperes peak-to-peak. In this circuit, there will always be a total current of 5.0 amperes, 2.5 amps to each transistor. Fig. 2-5 shows this basic circuit.

Let's get more out of it; if we bias each transistor to cutoff, that gets rid of the resting current that was doing no work. So, our rest current is zero. What will happen to the signal? We're still feeding in a sine wave, 180° out of phase, from the input transformer. The first half-cycle of the signal is positive-going on the top transistor and negative going on the bottom one. Remember, this is the *same* half-cycle of the signal. Fig. 2-8 shows the circuit now.

What will this do? Both of these transistors are npn's. Biased to cutoff, they will pass no current until the base is *forward-biased by the incoming signal.* During half-cycle No. 1, the top transistor is

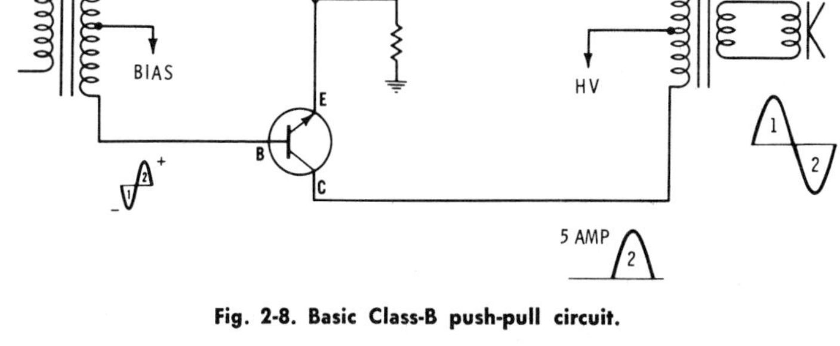

Fig. 2-8. Basic Class-B push-pull circuit.

biased on, and conducts. The bottom transistor is driven farther toward cutoff by the negative-going (reverse-biased) signal, and just sits there waiting. The top transistor is driven to full conduction (5 amperes) by the signal. This current flows through the primary.

On the next half-cycle, the same things happen in reverse. Now, the top transistor is biased off by the signal, and does nothing. The bottom transistor is biased on, and conducts its 5.0 ampere current, which also goes through the transformer primary. The two half-cycles in the primary come out in the secondary as the original sine-wave signal.

Now, what have we done? We've gone from a 5.0 ampere peak-to-peak current in the primary up to a 10.0 ampere peak-to-peak current, and increased the total output power by quite a lot. Each transistor can now conduct up to its maximum capacity, and this is all pushed through the primary winding. In fact, each transistor "pushes" current through from opposite ends. This circuit was originally called "push-push," to distinguish it from push-pull. Now, we just say "Class-B" and let it go at that.

Remember this push-push action. We're going to see a great many circuits using it, and some of them aren't going to look alike. However, they all make use of this basic Class-B reaction. If you remember the typical characteristics of the circuit, it will be easier to service. The one thing that distinguishes Class-B operation from the rest is the conduction of one transistor on one-half of the cycle, and the conduction of the other transistor on the following half. While each transistor is conducting, its opposite number is cut off. Transistors in Class-B, without an input signal, are always biased to cutoff.

THE OUTPUT-TRANSFORMERLESS CIRCUIT

Quite a while ago, while looking for a circuit that would be less expensive to build and still be efficient, the designers came up with a good one. This has all of the advantages of push-pull operation and all of its efficiency, yet has no output transformer at all. This has become perhaps the most popular and frequently used circuit in many kinds of transistor amplifiers. You'll find it in tiny phono amplifiers with a full 2 watts of power and in the monstrous 300-watt "Bass Amps" and pa systems too. This is called "output transformerless" or OTL for short. Since these are built in many different ways, they are generally known by their circuit design. Regardless of this, they all work in exactly the same way. You'll see examples of each type in later chapters.

Fig. 2-9 shows the basic circuit. This one is called a "stacked" circuit; it uses two transistors of the same type. They're actually

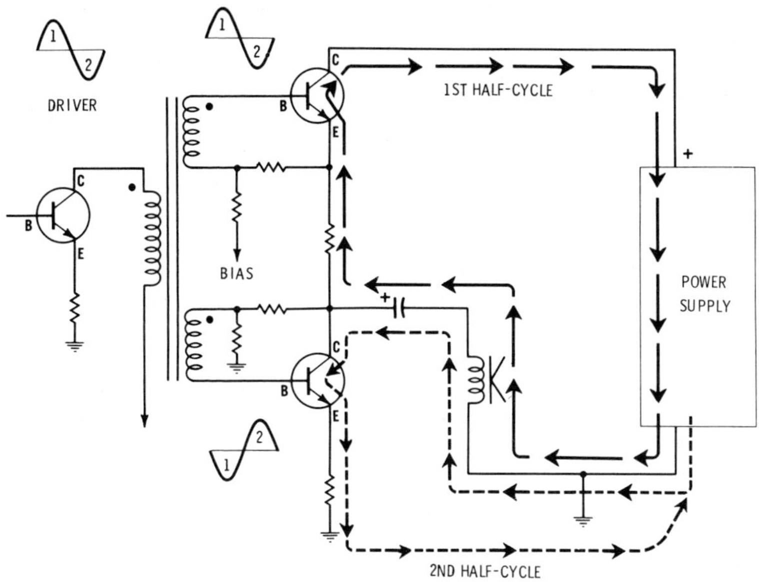

Fig. 2-9. Basic OTL circuit.

connected in series, with the emitter of the top transistor connected to the collector of the bottom one. The speaker is connected directly to the junction or center tap of the two, with a huge blocking capacitor between it and ground. (Remember this capacitor: it plays a very important part in the circuit action.)

Let's follow the signal through this circuit and see what happens. Since this is a Class-B circuit, both transistors are biased to cutoff; they draw no current at all unless they're getting an input signal. They're "resting." On the first half-cycle of the input signal, the positive-going half, the top transistor is driven into conduction. The bottom transistor is getting a negative-going signal at this time, so it is being driven deeper into cutoff and takes no current at all. The signal-current flows through the top transistor, through the big capacitor, then through the speaker voice coil to ground. The dashed line shows electron flow during this time. The current charges the big capacitor, and also makes the speaker voice coil move back and forth in step with the current.

On the second half-cycle, the signal changes polarity. It cuts off the top transistor, and makes the bottom one go into conduction. So, the second half of the cycle flows through bottom transistor, the speaker, to ground. But, where does the *voltage* come from to make the bottom transistor conduct? The top transistor is *between* its collector and the positive terminal of the power supply, and it is cut off completely by the bias plus the signal. It's an open circuit.

This has been one of the most difficult things to understand about this circuit, yet it's very simple once you know what happens. A great many persons have been puzzled by this. The voltage is there, all right, or the transistor couldn't conduct.

What happened during the first half-cycle while the top transistor was conducting? A big surge of current came from the power supply, up through the speaker voice coil, and the *capacitor*. So, the capacitor became *heavily charged*. Being a good capacitor (we hope!) no "dc" flowed through it, but the ac signal-current did. Since this was in the form of a dc pulse, the capacitor charged. In fact, that's all the top transistor does: charge this capacitor. The charging current has to come from somewhere, so it comes through the speaker, making the voice coil move at the same time. The polarity of the charge is just what we need, positive toward the collector of the bottom transistor.

When the second half-cycle arrives, and starts the bottom transistor conducting, it simply drains off the charge on the capacitor, which in effect is the same as drawing a pulse of signal-current through the speaker in the opposite direction. So, the voice coil is moved in the opposite direction, and we've got our full cycle of signal back again.

The whole action of this circuit, as you can see, is based on the charging and discharging of this big capacitor. You'll find some big ones in actual circuits; from 500 μF up to 1000 or even 2000 μF, electrolytics, of course.

In one of the prototypes of this circuit, the capacitor was left out. The speaker was connected directly to ground from the junction of the two transistors. However, this required the use of *two* dc voltage supplies, of opposite polarity, for the top and bottom transistors. So, this one wasn't popular long. You may run across it, in some of the older transistor amplifiers, but not too many. For reasons of economy, the single-voltage supply circuit with the charging capacitor is used in practically all amplifiers nowadays. It works just as well and is a lot easier to build and service.

MODIFICATIONS OF THE STANDARD OTL CIRCUIT

You're going to see a lot of different versions of the OTL circuit, of course. Because transistors come in two types, npn and pnp, you'll find the voltage polarity reversed. This makes absolutely no difference in the basic operation of the circuit. When you're checking the circuit, just remember which polarity of the bias makes each transistor *conduct*, and you've got it. As far as the actual output part of the circuit is concerned—that is, the two power transistors, the coupling capacitor and the speaker—you'll find that this is identical in all of them.

The modifications will all be in the *input* circuity: drivers, preamplifiers, phase inverters, and so on, plus several added features to ensure that the bias will be correct. The purpose of most of these will be to get rid of the input transformers that we showed, for simplicity, in the basic circuit. You'll find some apparently complicated circuits in this part. Just remember that they all *do* the same thing; they provide an input signal of the right polarity to the output transistors, and amplify it to the level needed to drive them to full output.

Bias circuitry will be more complex than the ones we've drawn so far. Same thing: all it has to do is apply the correct bias voltages on the output transistors, so that they will work exactly where we want them to. In several cases, you'll find dc feedback loops from output to input, in addition to, or as a part of, the ac feedback loops. The purpose of these is protection, in most circuits. They help to keep the big output transistors from self-destruction by thermal runaway.

CROSSOVER DISTORTION

Since this is a "thing" that happens in all Class-B amplifier circuits and can cause very severe distortion, let's take up crossover distortion right now. You will notice, in actual circuits, that the output transistors are NOT biased exactly to cutoff, as they were in the ideal circuits we have been drawing. This, like all other ideal circuits, is nice, but it won't work! The reason for this is complex, but what actually happens is this: Due to several things, we can't get

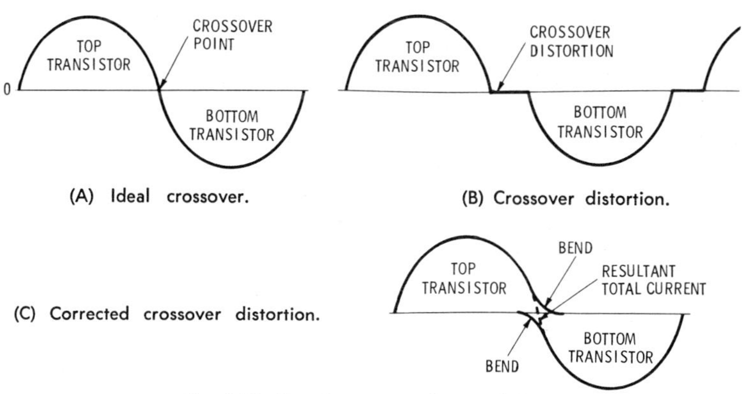

(A) Ideal crossover.

(B) Crossover distortion.

(C) Corrected crossover distortion.

Fig. 2-10. Class-B response characteristics.

a perfect, linear "transition" between the two transistors at the point where one stops conducting and the other one starts. Fig. 2-10A shows an ideal crossover with no distortion. If they *would* work this way, it would save a lot of trouble!

However, when we get down to using real transistors and parts, we get into trouble. The transition points on the curve do not quite fit, and we get a distortion of the sine-wave signal, as you can see in Fig. 2-10B. This is called "crossover distortion" (and other, ruder, things by engineers and technicians, on occasion). So, we must make adjustments to get rid of it. The typical symptom is that notch or zigzag in the curve at the point where it crosses the zero line from one transistor to the other. This is easy to see when viewed on an oscilloscope.

In practical circuits, transistors are not biased to full cutoff. Instead, they are biased to a point on the curve that permits just a little bit of current. This causes the curve to take on a slightly different shape at the crossover point, as you can see in Fig. 2-10C. The resultant of these two curves added together gives us what you see as the dotted line, and a nice linear transition point. By the way, this happens to be exactly the problem that tv designers faced when they built the first tv sets with damper tubes. They had the same kind of crossover distortion at the point where the damper stopped conducting and the horizontal-output tube started. This caused a white line to appear on the screen, right in the middle. The cure is the same; just change the bias a little until the resultant curve adds up to a straight crossover.

COMPLEMENTARY-SYMMETRY OUTPUT CIRCUITS

You've seen the stacked circuits with the same type of transistors, and a phase-inverter circuit or transformer that applies signals to the bases in the proper polarity, so that each transistor conducts at the right time. Now, let's take a look at a circuit that does its own phase inverting. This is another one of those ideal circuits in the first drawing, but later on you'll see plenty of the real things. (Just another way to get rid of that input transformer.)

Fig. 2-11 shows how this is done. What we do is take two transistors that are exactly alike, but of opposite polarity, one npn and one pnp. Now, we go back to our basic facts for a moment. An npn transistor will conduct only if a positive voltage is fed to its base; if a negative voltage is fed in, it cuts off. A pnp transistor conducts only if a negative voltage is fed to its base, and cuts off on positive voltage. That's the basis of this whole thing. Each transistor is the "complement" of the other; exactly like it in all characteristics, gain, and so on, but of the opposite polarity.

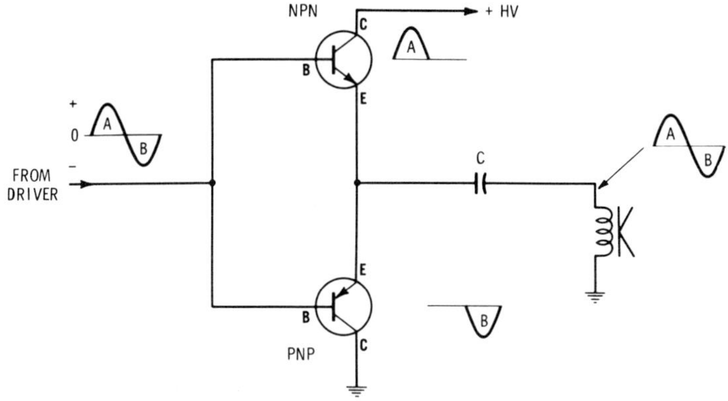

Fig. 2-11. Basic complementary-symmetry output circuit.

So, we simply tie the two bases together and feed an ac sine-wave signal to them. On the positive half-cycle (A), the top (npn) transistor conducts and the bottom transistor (pnp) ignores it. On the negative half-cycle, the bottom transistor goes to work, and the top transistor takes a rest. Once more, we get back our full sine-wave input signal, "recreated" from the two halves, across the speaker, and we're in business.

The same voltage and current reactions will apply in this circuit, just as with the stacked type. The dc voltage supply goes to the collector of the top transistor. When it conducts, the big capacitor is charged up; this supplies the voltage for the bottom transistor when its turn comes. It looks simple, but just wait until you see what is going to happen to it later on! Things will be added to it until you will think that it isn't the same circuit, but it will be. If you'll look closely enough, you'll be able to see the same old basic circuit in there, and the actions and reactions will still be the same.

THE QUASI-COMPLEMENTARY-SYMMETRY CIRCUIT

One modification that you're going to run into on many occasions will be the quasi-complementary-symmetry circuit. This means "Works just as if it were really a complementary-symmetry circuit." Fig. 2-12 shows one of these. The first thing you'll notice is that the output pair are of the same type, npn's. So, this much of it is the "stacked" circuit again. The complementary-symmetry part is in the two driver transistors. The top one is npn and the bottom one pnp. Note that the bases of these driver transistors are tied together, just as before.

As usual, the action is based on separating the input signal into positive and negative halves. The drivers have a sort of "double-

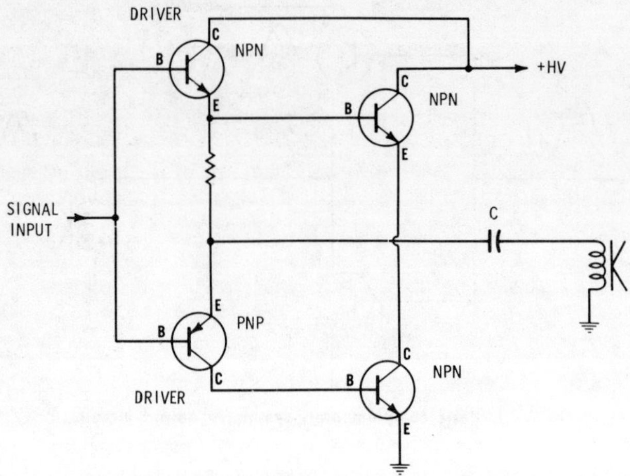

Fig. 2-12. Quasi-complementary-symmetry circuit.

action." When a positive signal voltage appears on the bases, the top driver feeds it right on through without changing its polarity. Notice that it is connected as an *emitter-follower*, which does not invert the signal. So, the signal appears on the base of the top output transistor as a positive voltage, and it conducts. This same positive half-cycle is applied to the base of the bottom driver. This one is direct-coupled to the base of the bottom output transistor, but as a common-emitter. So, the driver inverts the signal, and puts a negative pulse on the output base. This is reverse bias, and the bottom transistor stays cut off, while the top one is conducting.

On the negative half-cycle that follows, this will once more go straight through the top driver, and show up as a negative voltage on the base of the top output transistor. This stays cut off. The same negative half-cycle is inverted as it goes through the bottom driver, and comes out as positive voltage on the base of the bottom output transistor. This one conducts, and away we go again. The whole signal appears across the speaker, and everything is fine.

The drivers alone make up the complementary-symmetry part. With the versatility of transistors, you can fiddle them around in lots of ways, and come up with many different ways of doing the same thing.

There is one thing which must be remembered when selecting replacement transistors for use in these circuits. They must be matched pairs whether in the stacked circuits or the complementary-symmetry type. If the two transistors are not very accurately matched as to beta, voltages, etc., you'll get unequal amplification of the two halves of the input signal, and some real headaches.

BIAS AND COMPENSATION

The operating bias on these stages is very important. As we have said, it must keep the output transistors biased very close to ideal Class-B, but not exactly. In any case, the operating point must remain firmly fixed. Transistors have a nasty habit of reducing their internal (junction) resistance as they heat up. This makes the collector current go up, which in turn heats the junction even more, and so on and on. If this is allowed to go unchecked, they can go into thermal runaway and destroy themselves. (Anyone who believes the fable about "transistors run cool" is invited to touch a big power transistor after about five minutes' operation.)

Heat sinks are one way of holding this down. These are simply flat metal fins to give the transistor a bigger heat radiating area, and keep it as cool as possible. The transistor must be properly installed on the heat sink. In many cases the case (collector) must be electrically insulated from the heat sink, but still have the tightest possible thermal coupling for efficiency. They are usually insulated with very thin sheets of mica, coated with a special silicone grease for better heat transfer.

Compensation

Bias compensation is another method used to cope with some of the undesirable effects of heat. In Fig. 2-13 note the diode connected between the bases of the output transistors. Bias here is provided by a voltage divider from the high-voltage supply. The bases of the output transistors are about 1.0 volt apart. This is due to the voltage drop across the diode and the little resistor. If the collector current tries to rise because the transistors are getting hotter, the

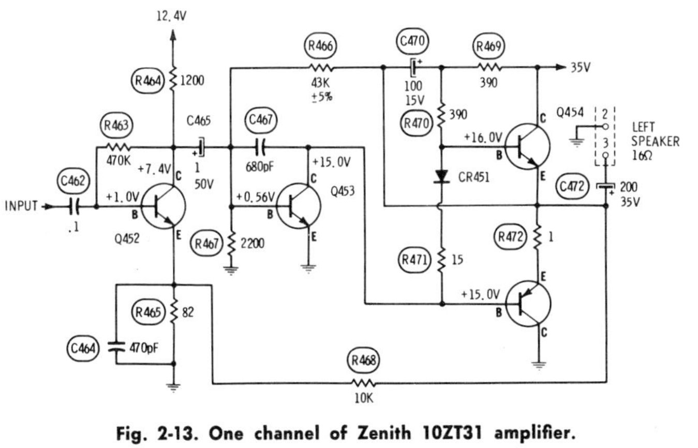

Fig. 2-13. One channel of Zenith 10ZT31 amplifier.

diode also warms up, its resistance changes, and it automatically changes the bias a little to hold current down.

The diode can do this because it is made of the same material as the transistors, germanium or silicon. So, its junction characteristics match those of the transistors. It is thermally coupled to the power transistors by mounting the diode tightly on the heat sinks. You can tell what type it is by checking the voltage drop across the diode(s). Silicon diodes will have about 0.7 volt, and germanium diodes about 0.2 volt minimum. The diode in Fig. 2-13 shows about a 1.0 volt drop including that across the resistor, so this is a silicon type, and so are the transistors.

The total voltage drop across the diodes, if more than one is used, can be found by checking the type, then counting diodes. Fig. 2-14 shows an output stage, of higher power, with two diodes in series. Here, there is a total voltage drop of 1.4 volts, or 0.7 volt each. So, these too are silicon. Incidentally, if one diode should short, you'd read only 0.7 volt base-to-base on the output transistors, and have some distortion. If both should short, you'd read zero volts base-to-base, and have quite a bit of distortion.

You may wonder about the diodes being "in series" with the input signal to the base of one transistor. If you'll check the polarity, you'll see how this can work. If a diode is forward-biased it acts like a closed switch, if the applied signal is not great enough to drive it to reverse bias. The diodes shown in Figs. 2-13 and 2-14 are forward-biased and have low ac resistance to any changes in driver cur-

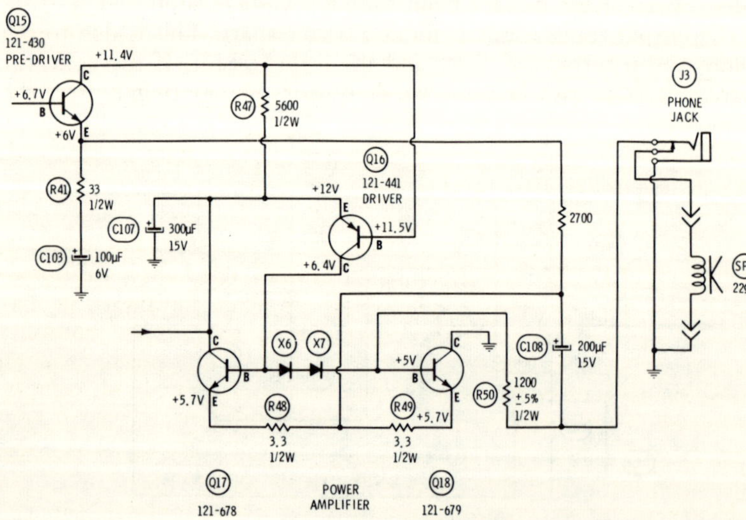

Fig. 2-14. Zenith 18ZT40Z3 amplifier.

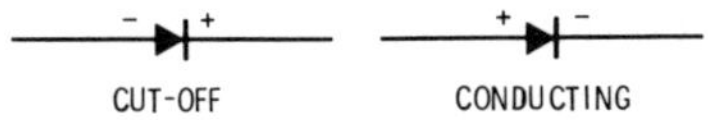

Fig. 2-15. Diode bias for cutoff and conduction.

rent. Fig. 2-15 shows the conducting polarity of all diodes when the voltage polarity is reversed (open circuit) or forward (closed circuit). (If you replace bias diodes and then have a terrific distortion, go back and recheck. You'll probably find the diodes connected backward.)

DC VOLTAGE DISTRIBUTION: THE KEY CLUES

In servicing these amplifiers, the dc voltage distribution is probably the fastest and easiest way of finding the cause of the trouble. There are some key voltages that will give you a great deal of information in a hurry. Check these first. With something like a "one channel dead" complaint in a stereo system, you can cross-check between the working channel and the faulty one, and pin it down quickly.

The most important single voltage in this circuit, aside from the dc supply voltage itself, is the almost even split of the supply voltage across the output transistors. This applies to all of the types of OTL circuits. The dc voltage of the supply will be almost evenly divided, and appear at the "midpoint," the common emitters in the complementary-symmetry circuit, the emitter-collector in the stacked circuit, etc.; in all cases, this is the point where the speaker and capacitor are connected. Note this voltage in the circuit of Fig. 2-13. With a supply of +35 volts, you should get +15.5 volts on the midpoint. This is not exactly half, but near enough. In Fig. 2-14, with a supply voltage of +12 volts, you get between 5 and 6 volts on the midpoint.

Don't try to pin this down to precisely half; not at first, anyhow. If you see a practically even split of the supply voltage, keep on looking. The output stage is working. If this channel is dead, this isn't it; test in the driver and preamp circuitry.

If you remember your circuit reactions, this key voltage will point out quite a few different faults almost instantly. For example, if the supply voltage is normal, but you find zero volts on the midpoint, the top transistor is *open.* This is probably the most frequently encountered trouble, especially in the smaller amplifier circuits. With this transistor open, there can be no current through it; so, the capacitor can't charge, the lower transistor has no voltage, and so on. There is, of course, always the possibility of a dead short here, from the bottom transistor being shorted collector-to-emitter. However, for some reason, this seems to be very rare.

If the midpoint voltage is quite a bit above normal, the top transistor is either shorted or leaking heavily. In a lot of cases, this will also blow the bottom transistor, especially if it happened to be a little underrated to start with. This kind of double-trouble is often found in the cheaper and smaller amplifiers, especially the imports, but is very scarce in the better quality units.

If the midpoint voltage is quite a bit low, you could have one of several things: leakage through the electrolytic capacitor and speaker to ground; collector-emitter leakage in the lower transistor, or a short somewhere in the dc feedback network, which usually takes off from this point and goes back to the drivers and preamps. This, too, is more common in the cheaper amplifiers, where the electrolytic capacitors are low-voltage types, more prone to high power-factor (excessive leakage) and shorting. This complaint will cause a symptom of very severe distortion and possible intermittents in the time *before* the channel goes completely out. When on a service call, ask the owner about this.

Direct-Coupled Circuitry

There is a high percentage of direct-coupling used in these circuits, as you can see from the two actual circuits we've seen so far (Figs. 2-13 and 2-14). The condition of the driver and preamplifier transistors will always have a definite effect on the operation of the output transistors. Leakage, shorts, or opens in any of these preceding circuits will upset the bias on the output transistors. So, if the output transistors will not work as they should, yet the transistors themselves test out all right, go back and check all dc voltages through the driver/preamp circuitry. Any kind of transistor trouble (which is the most common) in this circuitry will cause drastic changes in the dc voltages.

One good "quick check" for bias troubles is reading the dc voltages with *no-signal input*. In this condition, the output transistors should be drawing very little current. Cross-check between channels for bias voltages and currents. In one case, after replacement of the output transistors in one channel, these transistors were found to be running warm in resting condition. The transistors in the other channel were not; they stayed cool. There was also a definite increase in the distortion in the warm channel, as shown on the scope.

These symptoms pointed to a fault. Measuring the dc voltages showed that the warm channel had about 2.0 volts too much resting bias—forward, of course. This was traced to a driver transistor, direct coupled, of course, that had just a little bit too much leakage. Since this transistor's collector was directly connected to the output transistor base, this put too much forward bias on the output pair, and they drew current at rest, when they should have been cut off.

Very small dc-voltage changes in any of the early circuits will cause much greater changes in output circuits. These will be "multiplied" by the gains of the transistors in the circuit between the actual fault and the output stage.

One other (unfortunately somewhat common) fault in these circuits, which causes distortion and incorrect dc voltages, is a mismatch of the output power transistors. In all of these circuits, these two transistors must be a "matched pair" with exactly the same characteristics. Even though they are of opposite polarity, as in the complementary-symmetry circuit, characteristics, leakage, gain, beta, etc., must be precisely the same.

Fortunately for us, practically all of the major replacement-transistor makers have specially matched output pairs, in all sizes and types. Check the replacement listings for these. Above all, be sure that both of them are of the *same* type. One very odd case of distortion was finally cleared up by the discovery that someone had installed a silicon and a germanium transistor in the output pair. The difference in the required bias on these transistors gave a very peculiar looking output waveform, something like Fig. 2-16. Note the "flat" place along the baseline; this is a terrible crossover distortion, due to the mismatch.

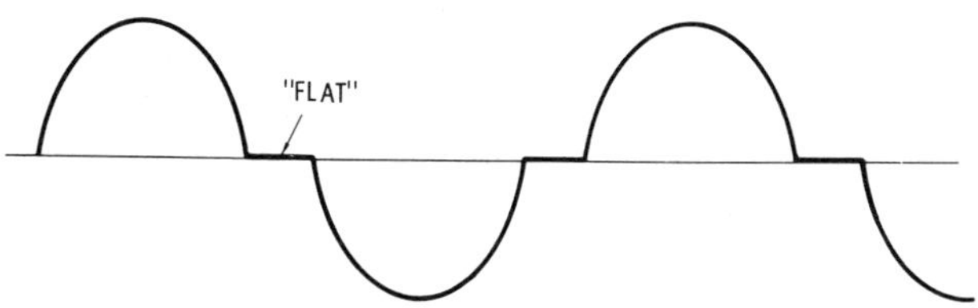

Fig. 2-16. Distortion caused by one germanium and one silicon transistor in output pair.

IMPEDANCE MATCHING: SPEAKER LOADING

There is one thing you cannot do, with this type of amplifier: change the speaker impedance. Only in specially designed pa-type amplifiers with a tapped-secondary output transformer can this be done. The typical transistor amplifier will be designed for a certain amount of power output into a specified load impedance, and that's that. Don't mess with it. You *can* make changes in the speakers, but *only* if you get back to the original impedance–two 8-ohm speakers in series substituted for one 16-ohm, etc.

Why is this? Let's see. Fig. 2-17 shows the output circuit and the load, which is the impedance of the speaker voice coil. The capacitor is so big that its series impedance, even at extremely low frequencis, is less than one ohm, so we can forget it. All we need is the speaker voice-coil impedance. This is the "load" across which we develop output *power*, by drawing current through it.

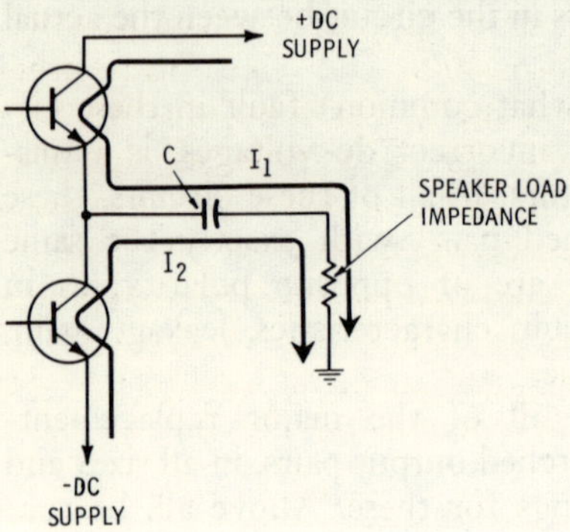

Fig. 2-17. Current through load impedance determines maximum power output.

Let's say that our amplifier is rated at 60 watts maximum power output, and has a 15-ohm speaker. Therefore, the output transistors will be pumping current through the 15-ohm load, to make this much power, by the I^2Z formula. Our current here would be 2.0 amperes. Squaring this and multiplying by 15 gives us 60 watts. This is maximum power output; the transistors have a safety factor of current to take care of peaks, and so on, and everything's "running cool."

What would happen if we took off the 15-ohm speaker and replaced it with an 8-ohm type? Our load impedance would be cut in half. Nothing else having changed, our output current would *double;* so, instead of 2.0 amperes, we're drawing 4.0 amperes. Using the I^2Z formula, we have $4^2 \times 8 = 16 \times 8 = 128$ watts. This, as you'll admit, is somewhat of an overload. The output transistors and the power supply will certainly think so.

These figures verify earlier statements about the different characteristics of the two major types of output circuits. The transformer-coupled Class-A types do not like open circuits; this causes trouble: they'll stand short circuits without damage. The OTL types, with no exceptions, *will* stand open-circuited outputs with no trouble, but will not stand up under short-circuited or reduced-impedance output.

So, you can go the other way, increasing the output load impedance, without any danger. Replacing an 8-ohm speaker with a 16-ohm will do no damage. The output power will be reduced by about half, but there will be no damage to components. This is a one-way exchange, though. DO NOT reduce the specified load impedance.

You have just seen why an 8-ohm speaker cannot be used instead of a 16-ohm, etc. These amplifiers are *designed* for a given power into a given load. To change this would require a complete *redesign* of the amplifier. Just changing output transistors to heavier-duty types won't do it. Seriously speaking, the design of a transistor audio amplifier of this type is an extremely hairy procedure, requiring a lot of high-level math, plus mountains of data on temperature coefficients, maximum load currents permissible, and many many other things. Redesign of these amplifiers in the service shop just isn't practical. We must confine ourselves to *replacement* of defective parts with *exact* duplicates. The only time we have any latitude at all in this area is when we replace power transistor types with almost exact duplicates having a higher breakdown voltage, for example, to attempt to get away from previous failures due to overload, insufficient power ratings in the original, and so on. If you've got one already designed, let it alone. It isn't practical to attempt to make major changes.

CHAPTER 3

Power Supplies

Just as it is in all other kinds of electronic equipment, the power supply is very important in transistorized amplifiers. It is common to all stages; it is also the place where a lot of the trouble is found. This is because of the fact that it is the hardest working section of the whole unit. The power supply should be the first thing tested in all service jobs. Unless it is in perfect working order, there's no point in making any more tests. (You'd be surprised how many men forget that.)

Power-supply circuits found in transistor equipment will all be "standard" circuitry—nothing new to the experienced technician. The same circuits have been used for many years in other types of amplifiers. The main differences will be the lower voltages and higher currents, as compared to tube-type amplifiers. There'll be only one that's even unusual, and this is just a different application, not a new circuit. This is the dual-voltage power supply, with both positive and negative voltage outputs.

Part sizes will be different. One thing that will look odd to the tube-type technician will be the filter capacitors. Some of these will be 1000, 2000, or even 4000 microfarads, far larger than comparable units in tube gear. One big amplifier uses two 4000-μF units in parallel, for a total of 8000 μF, as an input filter. This is due to the sensitivity of transistor circuits to even a very little ripple or hum. They need absolutely "pure dc." So, ripple tolerances in the power supply will be very, very small, preferably zero.

Voltage regulators will be found in many of the better power supplies. These are normally used to supply the preamplifier and driver stages, etc., rather than the power output stages. The idea behind this is to provide a fixed voltage source for bias voltages, etc., to prevent any shift of operating points. Since a change of only a few tenths of a volt can have serious consequences in transistor amplifier stages, the power supply should have very good regulation.

Some regulator systems will be as simple as a single zener diode across the voltage supply. This clamps the dc voltage to a given maximum level. Of course, this type of regulator can only hold the voltage below a certain value; it can't bring it up if it falls below the desired level. Other, more complex regulators, by the use of a transistor regulator in series with the main supply, will hold the voltage from shifting either up or down. The action of these circuits is so fast that they also help hold down the ripple. They'll be found labelled "Capacitance Multipliers" or "Electronic Filters" in some units. They actually "regulate out" the ripple, thus in effect multiplying the size of the main filter capacitor.

Transistor audio amplifiers come in all sizes, from the very small to the very large. Power supplies vary in similar fashion; they will run from about 2 watts up to 500 or even 1000 watts—enough to run a small broadcast radio station. However, they're all basically the same in operation, and are easy to service.

THE "LITTLE BITTY" ONES

There are a lot of very small transistor audio amplifiers. They're used in small mono and stereo record players and many other applications. Stereo players simply use two identical units with a common power supply. The whole amplifier, power supply, and everything except the speaker and controls is built on a tiny PC board, usually about 2″ wide and 3″ long. In the typical mini-amp, the board is mounted on a sheet metal bracket; the other half of this holds the volume and tone controls, and provides for fastening to the cabinet.

To save space and parts, even these mini-amps use the two-transistor output-transformerless circuit. The transistors used are generally the TO-5 case type, general-purpose audio, used for drivers in the bigger amplifiers. Power output on these runs from 1 watt up to perhaps 4 or 5 watts, so they don't need a very big power supply.

Fig. 3-1 shows a typical mini-amplifier, with three transistors. Note the power supply circuit; you'll find this used in a great many amplifier systems and almost universally in the record player amplifiers, both mono and stereo. In this circuit, the "power transformer" is the phonograph motor. A secondary winding is added to provide the necessary ac voltage, in this case 12 volts. In another common version of this circuit, the low-voltage ac is taken from a tap on one of the motor windings. This, of course, does not offer the isolation of the separate secondary winding; there is a shock hazard, so watch out for it.

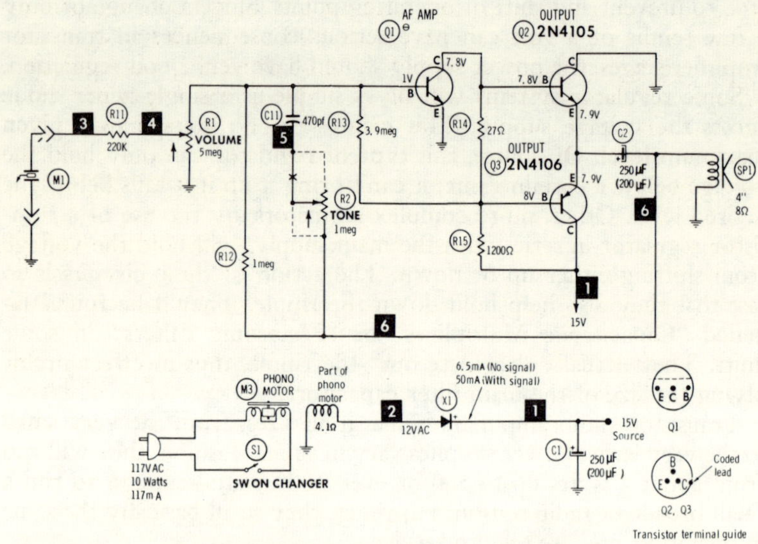

Fig. 3-1. Three-transistor, miniature amplifier.

The rectifier is a simple half-wave circuit, using a silicon diode. Filtering is provided by a single 250-μF electrolytic capacitor. Since the current drain is low, only 50 mA in this unit, no filter choke or extra smoothing is needed. This is the full-load current, dropping to no more than 6.5 mA at no-signal input.

There's not much you can say about such a simple unit. If the voltage is zero, check the ac input, with the motor running. If this is the normal 12 volts, but there is no dc voltage output, the rectifier is open. If the rectifier should short (more common), this will apply raw ac to the big electrolytic capacitor, which is a very good ac short circuit. The phono motor will probably get very hot due to the overload, and will run slow. If left on long enough, it can burn out. In replacing one of these motors, be sure that you get an exact duplicate; while the "motor" part is all the same, there are several different secondaries in use. Be sure to get one with the correct low-voltage ac supply for this amplifier.

If the dc voltage is low, disconnect the supply leads to the output transistors. Recheck; if the voltage comes back up, one of the output transistors could be shorted. The motor will probably be getting warm; the rectifier may be hot, too. If there is no sign of overheating but the output dc voltage is down to about half-normal, the input filter capacitor could be open. Since this is the standard "reservoir-capacitor" circuit, if it is open there will be no place to store the energy delivered from the rectifier. The load will draw the voltage down as fast as it builds up. To check for this, bridge a new capacitor across the suspected unit (with power OFF, as in all transistor circuits) and turn it back on. If the original capacitor was open, the voltage will come back up. However, if the original capacitor had a high power-factor (high leakage) this test won't be conclusive. Best check is to unhook the original capacitor and clip in a new one for test. With these low-voltage electrolytics, high power-factor is a common complaint, so be on the lookout for it.

Measurement of the input *power* is perhaps the fastest way of checking these amplifiers, especially for power supply troubles or shorts. The normal drain of this one is only 10 watts. If you plug it in and find a reading of 30 or 40 watts, you very definitely have a short circuit somewhere. If the input drain shows only the normal drain of the phono motor, about 5 watts, the amplifier is dead—probably an open circuit, open filter capacitor, etc. This test, and the equipment needed, will be discussed in detail in the next chapter.

THE "LINE-CONNECTED HALF-WAVE" POWER SUPPLY

Even our old friend from ac/dc radio-tv circuits, the straight line-connected half-wave power supply, will be found in some of these amplifiers. These will be mostly in the later models. Early versions had only low-voltage transistors to work with, so they had to have a 12- to 15-volt dc supply, etc.

However, high collector-voltage transistors were developed a few years ago for use in video output stages in tv sets. This allowed the use of higher voltages and greater power. Fig. 3-2 shows a power supply of this type. The dc voltage comes out as +126 volts. The audio output transistor used in this amplifier works with +118 volts

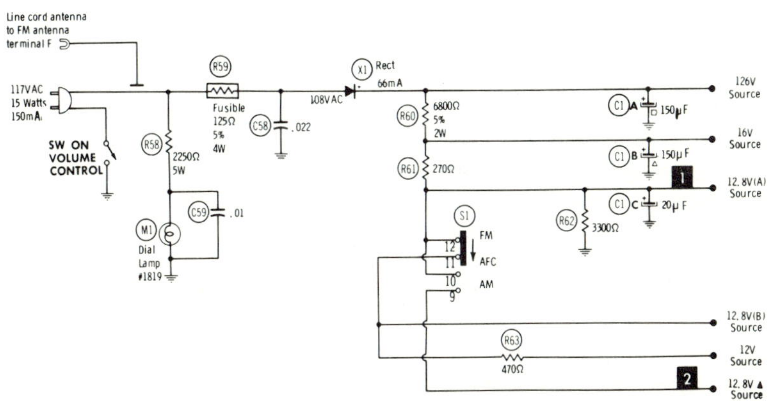

Fig. 3-2. Line-connected, half-wave power supply.

on the collector. This collector breakdown voltage is something that you must check when making replacements. There are a lot of similar transistors, but with much lower breakdown voltages.

Lower dc voltages for preamplifiers and drivers are taken from a voltage divider. In this amplifier, +16 volts and +12.6 volts are used. Tests in this circuit are the same as in all the rest. Input power drain of this amplifier is only 15 watts. It uses a single transformer-coupled output transistor, in Class-A, which accounts for the higher input power, with no signal.

If the dc voltage at the rectifier output is down to about half normal, 50 to 60 volts or less, but there is no sign of anything overheating, check the filter capacitor for an open circuit or high power-factor, by replacement. This uses a single silicon diode rectifier; these often short, but seldom open. Open rectifiers are very rare, but certainly not impossible.

MEDIUM-POWER AMPLIFIERS

In a good many of the medium-powered amplifiers—meaning those from 10 to 30 or 40 watts total—low dc voltages are used. So, they'll have step-down power transformers. From this point on up to the very large types, the power transformer is a necessity, to get the low voltages and high currents needed. A great many of these amplifiers will be found in "stereo theater," home record players of the larger types, especially stereo systems, and in combination with am/fm radios and tv sets. Output power of 40 or 50 watts (total, 25 watts per channel) is common lately. This used to be, and still is, sufficient power for a good sized pa system.

The power supply circuit of Fig. 3-3 is a typical example of this size. The secondary supplies 27 volts ac, center tapped, to a full-wave

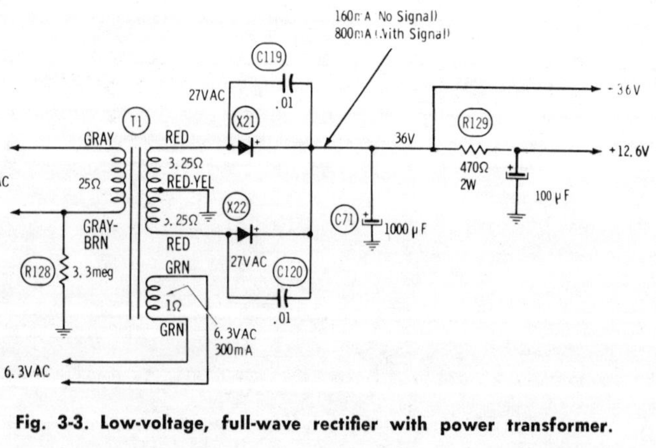

Fig. 3-3. Low-voltage, full-wave rectifier with power transformer.

rectifier circuit. Silicon diodes are used; in fact, these now are universally used. This particular amplifier is used in a good-sized home music system, so more filtering is needed. This uses a π-section filter, with a 1000 μF electolytic input filter, and a 470-ohm, 2-watt resistor as the choke. While the output filter capacitor shows only 100 μF, there are quite a few others used on this line elsewhere in the circuit. So, the total for the output filter is considerably greater than 100 μF. Most of the other circuits are isolated with small resistors, and another 100 μF capacitor is used to prevent feedback and hold the ripple down to the absolute minimum.

Low-voltage output to the output stages is +36 volts. This feeds both channels of the stereo amplifier. Other stages are supplied by a dropped voltage of 12.6 volts. The power amplifier draws 160 mA with no signal, and 800 mA with signal input.

Input power drain is once again the quickest check for possible trouble. This amplifier and its associated am/fm radio draws only 13 watts, with the record-changer not running, with no signal. If there is an internal short anywhere, or high leakage, this will shoot up to a value big enough for you to notice instantly. Shorted rectifiers, leaky or shorted filter capacitors, or trouble in either output stage will cause the no-signal power drain to rise.

To isolate the cause of the trouble, after initial tests have shown an overload, disconnect the two output stages from the dc power supply. If the power input drops back to normal or quite a bit below normal, there you are. Dc voltages can be measured; if these are slightly high because of lack of loading, the dc power supply is all right. Reconnect the output stages one at a time, watching the power input. This will tell you which channel is at fault.

Another test method that can be very useful here is breaking the dc supply circuit at the rectifier output and inserting a dc milliammeter. Read the no-signal drain first, for each channel. This should be roughly half the maximum no-signal current given on the schematic, or about 75 mA in this amplifier. Overloads will show up instantly. Also, a low current drain on a given channel is a very useful indicator, too.

If one channel draws practically no current at all, this is very often due to an *open* output transistor, the top transistor in a stacked or complementary-symmetry stage. When this transistor opens, the whole stage can get no current since it all flows through this one. A double-check for this condition is a dc voltage reading at the midpoint, where the speaker is connected. This should normally be about half of the supply voltage. If the top transistor is open, the reading will be either zero or very low, indicating trouble.

If the amplifier is dead, and the 470-ohm choke resistor is open and very badly burned, do NOT replace it and immediately turn the power on. Something blew this resistor and whatever it is, is still in there. Find the short circuit first, and fix it. A very useful test method for such cases is given in the next chapter.

FULL-WAVE BRIDGE CIRCUITS

Low-Voltage Power Supply

Fig. 3-4 shows a power-supply circuit that is almost exactly the same as the one in Fig. 3-3, but uses a full-wave bridge rectifier. You'll find quite a few of these, since they are slightly more efficient than the two-diode full-wave rectifier and give a lower ripple. Voltages and currents in this amplifier are in the same range as the preceding one. Note the 140-mA no-signal current and the 520 mA maximum. So, this unit has about the same power output as the other.

All tests and measurements given for the preceding circuit will apply to this one. Note the big 2000 μF input filter capacitor. You will find bridge rectifiers made up of separate silicon diodes mounted on terminal strips, or encased in a single unit with four leads coming out. A shorted diode in a bridge will ordinarily reduce the output voltage by at least half. It will also blow the fuse or trip the circuit breaker, and the input power reading will go far above normal, usually to 100 or 150 watts, or even more. If the amplifier is *not* protected by a fuse or circuit breaker, a shorted diode in the bridge can burn up the power transformer in a comparatively short time. For a quick-check on the condition of the transformer, disconnect the two leads to the bridge, breaking the dc circuit, and apply ac power. A good transformer will show practically no input power reading under these conditions of no load. If you get a reading of quite a few watts, sometimes only 30 or 40 watts, and you're sure that there *is* no load on the transformer, it has shorted turns and must be replaced. If the short is a bad one, you'll find no-load readings of 100 to 200 watts or more (and it will probably smell!).

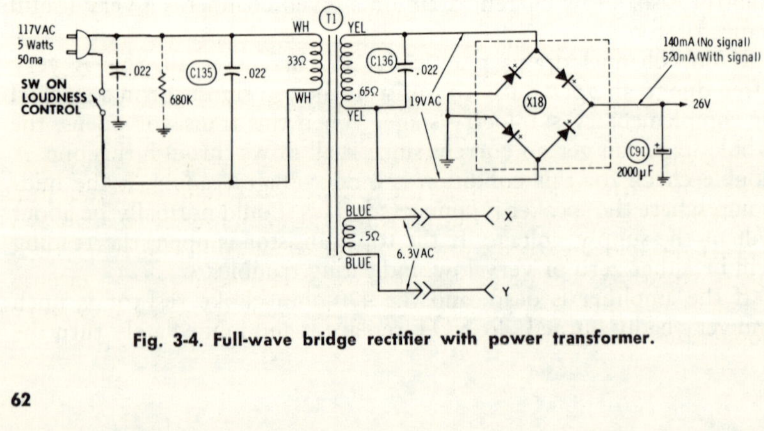

Fig. 3-4. Full-wave bridge rectifier with power transformer.

Dual-Voltage Power Supply

In certain circuits, you'll find dual-voltage power supplies with equal positive and negative voltages. These are needed for the "floating" type of OTL output stages, where equal and opposite voltages are fed to the top and bottom transistors. Fig. 3-5 shows how this can be done by using a full-wave bridge rectifier and a power transformer with a center-tapped secondary.

The center-tap is the common and is grounded. Each half of the bridge acts as a regular two-diode, half-wave rectifier. One is a common-cathode for the positive supply, and the other a common-anode for the negative supply. Each half has its own filter system, since the polarity is different.

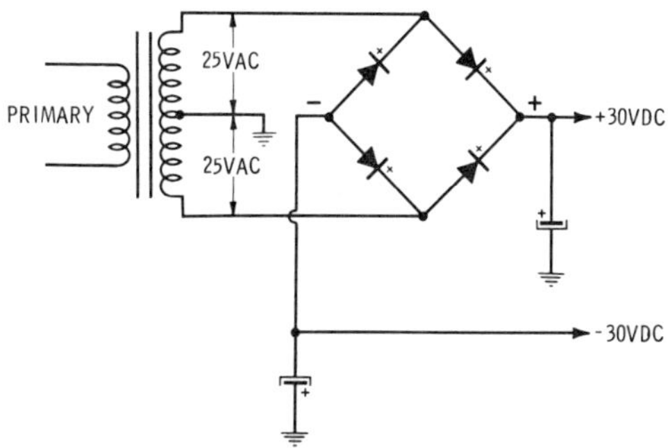

Fig. 3-5. Dual-voltage supply from full-wave bridge rectifier.

Other loads may be tapped off this supply as needed for preamplifiers and driver stages. This is a relatively light loading, and these stages commonly use heavy-duty rectifiers. Therefore, this additional load won't unbalance the power-stage output voltage supply.

Troubleshooting is the same as with the others. Load-shedding in case of overloads and shorts will show up the location of defective capacitors, rectifiers, and so on. Input and output current and power testing will work just the same to help pin down the bad part.

VOLTAGE REGULATORS

In many of the better amplifiers, you'll find voltage regulator circuits. These can be as simple as the zener diode circuit shown in Fig. 3-6. The dc power supply output is fed through a series resistor, and a zener diode is shunted to ground at the other end. The zener voltage of the diode will be of whatever value is needed for the circuit. You'll find a lot of 12-volt zeners in use. This seems to be a very popular low voltage value.

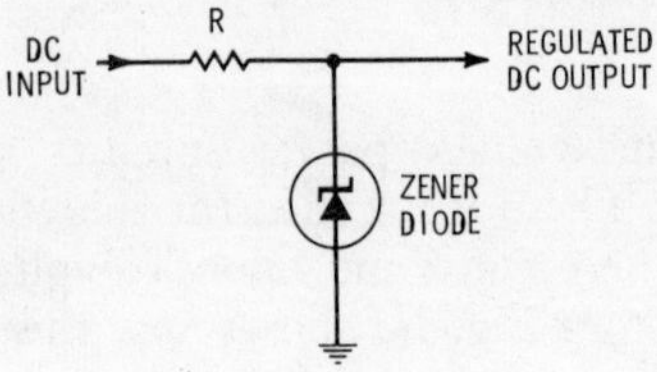

Fig. 3-6. Simple zener-diode voltage regulator.

Zener Regulators

The action of these regulators is simple; when the supply voltage rises above the zener voltage, the zener diode starts to conduct heavily. This causes a voltage drop across the series resistor, holding the voltage down to the zener value. Size of the resistor and the wattage of the zener are chosen so that they will stand up under the normal loading. The wattage of the zener is determined by the zener current at full conduction. Like diodes, they come in quite a few different wattage ratings. When making replacements, be sure to get an exact duplicate or to at least get one which will stand up in this particular circuit. Don't use a ½-watt zener for replacement of a 1-watt, for instance.

Voltage-checking of these is simple. If the dc voltage is well below normal, disconnect the zener. If the voltage jumps up to a value well above normal, the zener is leaking. Replace it. If the voltage is above normal at all times, the zener diode is probably open. Disconnect it; if this has no effect on the voltages, this confirms the diagnosis. If the dc voltage is low at the zener diode and disconnecting the zener has no effect on it, there is an overload somewhere in the circuits being supplied by the zener. Disconnect these one at a time until you find out which one is pulling the voltage down.

Transistor Voltage Regulators

The collector-emitter junction resistance of a transistor can be used as a "variable resistor." If we vary the base bias, the current through this junction changes. This means that the resistance has changed. By connecting the collector-emitter junction of a transistor in series with the dc power supply and clamping the base, we can use this as a voltage regulator. Fig. 3-7 shows a typical circuit as used in Zenith amplifiers.

The dc output current flows through the collector-emitter junction of the regulator transistor. The base of this transistor is held at a nearly constant voltage by connecting a large capacitor from this point to ground. When this capacitor is fully charged, it acts as a reservoir, or "battery"; it tends to keep the base at a constant voltage.

When a ripple pulse (from the rectifier) appears on the collector, it would normally go through the transistor and appear as a ripple

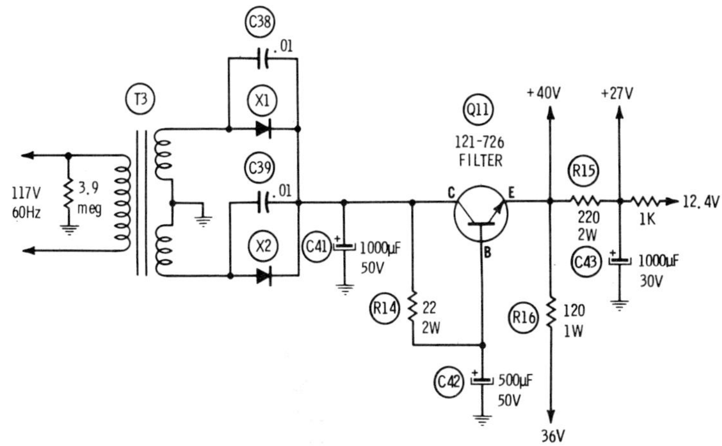

Fig. 3-7. Voltage regulator or "electronic filter."

on the dc output. However, this same ripple pulse is also fed to the base of the transistor, where it is of the right polarity to "cancel out" the variations in the collector current. So, this action smooths out any ripple in the dc output. If the collector-emitter resistance were *fixed*, a change in the supply current, or ripple, would cause ripple in the output. By cancelling out this ripple with a ripple pulse on the base, the transistor actually serves as a "filter." Since the base and collector voltages (from the ripple) are always in-phase, the base is able to "reduce the resistance" of the collector-emitter junction and smooth out any variations in the output.

The big capacitor on the base acts as a clamp. In effect, this capacitance is *multiplied* by the current-gain (beta) of the control transistor. So, this circuit is sometimes called an "electronic filter" and sometimes a "capacitance multiplier." Either one is correct. If this transistor has a beta of 100, the "effective capacitance" at the output would be 500 μF $\times$ 100, or the equivalent in filtering efficiency of a capacitor of 50,000 microfarads.

Series Regulators

The circuit of Fig. 3-8 looks a good deal like the preceding one, and indeed works in the same way. The collector-emitter resistance of the transistor is the controlling element. However, the base voltage here is controlled by a sampling circuit. A bleeder resistor or voltage divider is connected across the regulated output, and the base is returned to a suitable point, usually to a variable resistor.

This variable resistor is set to a point that will let the base hold the collector-emitter junction at the desired resistance, to give the dc voltage output needed at the regulator output. Now, if the dc output

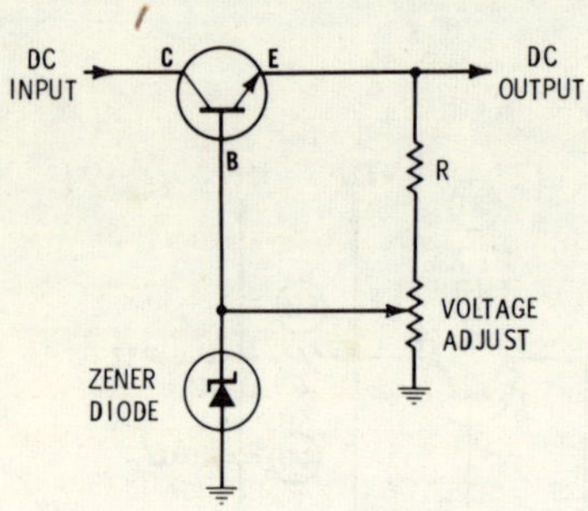

Fig. 3-8. Basic transistor voltage-regulator circuit.

voltage goes up, this causes a change in the transistor base voltage, which increases the collector-emitter resistance, pulling the voltage back down. The opposite happens for a drop in output voltage; the base makes the junction resistance lower, and the voltage is pulled back up again.

Fig. 3-8 is only the basic circuit of this type of regulator. You'll find many modifications, such as zener diodes used to clamp the base, and "error-amplifiers" connected between the output voltage divider and the control transistor base. This makes them more sensitive to very slight changes. You'll find some of these regulators with resistors shunted across the collector-emitter junction. This is done to carry a part of the load-current, and allow the use of a lower-power transistor in the series-regulator position. The regulation is slightly poorer, but adequate. Just remember the basic circuit action of Fig. 3-8; they all use this principle.

Voltage regulators are easy to test. If there is normal dc voltage input from the rectifier and filter, but no regulated dc output, check the regulator transistor, its dc voltages, and the loads. Disconnecting the load from the regulator output should let the voltage rise to the regulated level. If it does, and if it will hold pretty close to this level at varying ac input voltages, the regulator is working; there is a short in the load. This symptom could also be due to an open regulator transistor, or to a faulty error-amplifier transistor, zener diode, etc., which has biased the control transistor to cutoff. If the control transistor has the shunt resistor from collector to emitter, an open transistor here will let the voltage rise above normal. There will be no regulation at all, of course. The output voltage will go up and down as the ac input voltage is varied.

PARTS REPLACEMENT IN POWER SUPPLIES

Rectifiers

The only precaution necessary when selecting replacements for defective silicon-diode rectifiers is to be sure that the new one has an ample safety factor of both current and voltage rating. Normally, in low-voltage circuits you won't have to worry at all about the voltage rating, since practically all silicon diodes have high peak-inverse voltage (piv) ratings.

The current rating is very important. In the lower power amplifiers, where maximum current demands run from about 150 to say 800 mA, a 1.0 ampere (1000 mA) rectifier will do very well. However, for safety it would be better to use one with a 1.5 or 2.0 ampere rating. The higher-rated units will run just a little bit cooler, and thus have a far longer service life. Stock silicon diodes in the lower current ratings, say up to about 3.0 amperes, do not use heat sinks. In the big amplifiers, where currents can run up to 8 or 10 amperes, you'll probably find the rectifiers heat-sink mounted. For these, take all of the same precautions you would when replacing high-power transistors. Check the maximum-current ratings given on the schematic. If you are unlucky enough not to have an exact replacement, pick a rectifier that you're pretty sure will take it, install it temporarily, and take a reading of the actual current, with full signal input. In Class-B stages, this will sometimes be 4 or 5 times the resting current, so don't forget this.

Filter Capacitors

Shorted or leaky filter capacitors will make up a good part of your parts-replacement problems in power supplies. When selecting a relacement, you do not necessarily have to have an exact duplicate. Filter capacitors aren't critical in one direction–you can always use a *bigger* capacitor, but it isn't a good idea to use a smaller one. In other words, if you needed a 1500 μF electrolytic, you could use a 2000 μF, but a 1000 μF might leave just a little too much ripple. Actual ripple measurements with a scope should always be made whenever electrolytic capacitors have been replaced. If possible, get the ripple down as close to zero as you can by using larger filters if need be.

The low-voltage electrolytics seem to be prone to leakage troubles, especially in the cheaper import amplifiers. When choosing a replacement capacitor, always use a liberal safety factor in the *working-voltage* rating. For example, if the original was a 25-volt type, and your working voltage in the circuit is say 16 volts, use a 50-volt replacement. It will have quite a lower leakage, and will last longer.

Many of the capacitors in the bigger amplifiers are of the twist-lock or metal-can types. Some are single units; others are multiple-section types. If any one section of a multiple electrolytic fails, always replace the whole thing. Do NOT shunt a replacement unit across a faulty section, even if you clip the lead to the bad unit. I have even found replacement capacitors soldered right across defective units without even cutting the bad section out! This works for a short time, but in numerous cases the defective (open) section has healed itself, then shorted. Whatever conditions prevail inside the can of a multiple electrolytic to cause one section to fail will affect *all sections* of that unit, so replace the faulty one.

When replacing can-type capacitors, always take special pains to solder the lugs very tightly to the chassis. This is a common-ground return path for all currents circulating through this capacitor (signal currents, most especially). So, if the ground connection is not firmly made, you can have mysterious troubles. You'll notice that in the better amplifiers, all four lugs of the can are often soldered to the chassis. Your little "PC-board" type iron won't do this; a 100-watt solder-gun is needed for this kind of work.

Replacement Power Transformers

There was a time, not too long ago at that, when the low-voltage, high-current replacement power transformers were pretty hard to find. Now, however, there are ample listings of transformers of any desired size, shape, and type in replacement transformer manufacturer's catalogues. You'll even be able to get exact duplicate types, by checking under the original maker's name and part number.

One thing that is very important in replacement power-transformer selection is the current rating of the secondary. You can have exactly the right voltage rating, but if the secondary won't handle the needed amount of current, your new transformer is going to go up in smoke in a very short time. Check the current drawn from the power transformer *after* the replacement has been installed. This will give you a very good double-check on how the new one will stand up.

If possible, check the maximum load current shown on the schematic. Say that this is 800 mA, as in one of the amplifiers mentioned earlier. For a good safety factor, get a transformer with the same secondary voltage, and at least a 2.0 ampere maximum-current rating. Under an 800-mA load, this transformer will run a great deal cooler (and longer) than another transformer with the same voltage but only a 1.0-ampere rating. "Watch out for those watts!" They can sneak upon on you. Thirty five volts at 1.0 ampere doesn't sound like too much, but it's equal to 35 watts of power, and that's quite a lot.

If there is no information available on the maximum current drawn by an amplifier, and the original power transformer has burned up, it would probably be a good idea to try to fire up the amplifier with a substitute power supply, and get some figures. You can get an idea of about what the maximum dc voltage was in the circuit by checking the working-voltage rating on the electrolytic filter capacitors. If this is, say 25 volts, you can be pretty sure that the original transformer had a dc voltage output of not more than 20 volts. Applying the correction-factor, this would mean that the raw ac secondary output would be somewhere in the vicinity of 16 volts, rms. By checking the type of rectifier circuit used, you can tell whether the original transformer had a tapped secondary or not. Single-polarity power supplies with a full-wave rectifier need a center tap. Single-voltage power supplies with a bridge rectifier do not. In the first case, measure the ac voltage between one end of the secondary and center tap; in the second, read the ac voltage from end to end of the secondary.

In the following chapter, we'll give you a very useful test method for making sure of your power supply without the possibility of doing further damage to the amplifier circuitry. We'll also give you some hints about a few very useful test instruments for use in the area of the power supply.

CHAPTER 4

Test Instruments and Test Methods

Now, you've seen the circuits used in this type of audio amplifier, and how they work—when they're working. What you need now is to know how they work when they're *not* working! In other words, you've got the normal reactions; what you need is the *fault-reactions*. This is the most important part. You must learn to recognize fault-reactions—what they mean and how they can point to the cause of the trouble.

For this, you must have one important thing, the right test equipment. Since we are working with things we can neither see, hear, smell, feel, or taste, we must have instruments, as extensions of our human senses, to get into these circuits and give us information about what's going on. That is all a test instrument does; converts these "invisible reactions" into something that can be "read"; a meter reading, a scope pattern, etc.

You will be surprised at how few pieces of test equipment are needed, compared to other kinds of electronic servicing. You can test and repair any transistorized audio amplifier with only four instruments: a vtvm, an oscilloscope, an audio signal generator, and a transistor tester. There are a few pieces of auxiliary equipment that will make your work easier; these will be taken up as we get to them. However, for ordinary work in transistor audio circuits, the four instruments just mentioned are all you will need.

Certain instruments are ideal for one use; others cannot do this thing, but something else. For example, you can read ac or dc voltages very quickly and accurately with a vtvm. You can't tell *what* you're reading with the ac voltmeter, though. You can with the scope. You can read either ac or dc voltages with a scope (of a certain kind), but it's easier with the vtvm. So, use the voltmeter for its "specialty" and the scope for its specialty, and you will get along a lot better. There is only one qualification for any piece of test equipment. "Will it get this job done quicker and easier?" If the answer is yes, then that is the one to use.

All of the test equipment described in this chapter has passed this test. All tests and test methods are based on actual bench-work on real transistorized equipment. The equipment has been used on enough different units to get a statistically accurate sampling of its ability to do the job. Comparison tests between different ways of doing the same thing have shown which one is the fastest, and this is the one I'll give you. All we really have to sell is time. If we can save that on each job, we make more money. If a test instrument will save you time, it will pay for itself in a short time; if it won't, you've wasted your money on it.

THE "BIG FOUR" TEST INSTRUMENTS

The VTVM

You'll be able to do most of this work with the volt-ohmmeter. From much experience, I prefer a very high impedance type: the vacuum-tube voltmeter, transistorized voltmeter, or FET voltmeter (all of these are "volt-*ohmmeters*," by the way). To save time, I'll just say "vtvm" from now on, and you can pick the type that you like best.

These instruments are very sensitive, and they are accurate within the tolerances we have to work with. Best of all, they are practically burnout-proof due to their design, even the transistorized and FET types. With their tremendously high input impedance—up to 15 megohms in some FET types—they load the circuits very little, so that the voltage you read is very near to the actual voltage.

As you have seen, dc voltages on transistor elements are a very valuable clue to the performance of the stage. Any trouble will show up as incorrect dc voltages. With the proper interpretation of what these incorrect voltages mean, you can find trouble fast. Current readings can be made with a dc voltmeter, simply by reading voltage drops across known-value resistors; you do not have to break the circuit and insert the milliammeter (or ammeter).

The Oscilloscope

The oscilloscope is the only instrument in the shop that is really practical for signal-tracing. You're going to hear a lot about this before we get through. It will not only show you the signal, and its relative level, but it will instantly show up any distortion, variation of gain, and many other things. It is much easier than any other instrument for this kind of work. Like the vtvm, it, too, is completely burnout-proof. If you connect it to too high a voltage, the trace simply goes off the screen; no harm done. Like the vtvm, it also has a very high input impedance, which can be made even higher by the use of a low-capacity probe.

For most audio work, you'll use the simple direct probe. This will save you the trouble of having to make a mental correction for the multiplication factor of the low-capacity probe, too. If your scope has voltage calibrations on its vertical attenuator, it can be used for making gain measurements in terms of the actual peak-to-peak voltage. However, for most of this work, we won't have to bother with that at all. All we need to know is the value of a signal-voltage *relative* to a reference voltage; for example, the output voltage of a given stage relative to the input voltage. Comparison of these two will tell us instantly whether the stage has a gain or a loss, and if so, how much.

Wideband scopes are not necessary for this, although they will work very nicely. Many shops have older scopes, set back on the shelf because of their low bandwidth, not suitable for color-tv work, etc. These will be just right for audio work, where the highest frequency we'll run into will be perhaps 20,000 Hz. Even if they do not have internal peak-to-peak calibration, an external calibrator can be used, although we seldom need *exact* signal-voltage readings.

The Audio Signal Generator

This is exactly what it says: an instrument that can generate a signal in the audio range anywhere between 10 Hz and 20,000 Hz. It doesn't even have to be variable, for most tests. It will be more useful if it is, but it doesn't absolutely have to be. Any one of several types of audio generator on the market will give you a good sine-wave audio signal, controllable in level, and many of them will also make square waves, which can be very useful in certain cases.

As a matter of fact, any source of audio signals can be used: a record player or tape-deck, or even the audio output of a small radio—anything at all that will furnish an audio signal to feed through the amplifier under test. As long as you know what your test signal looks like on the scope, you can follow it all the way through the entire circuit.

The Transistor Tester

There are a lot of transistor testers on the market now. Those made by the better instrument-makers are fine instruments, and will be a lot of help to you. Practically every one of them will go right into the circuit and read the quality of a transistor without any need to disconnect leads. There are exceptions to this, of course. In the direct-coupled circuits, you can't make this type of test. There are too many things shunted across the transistor you're trying to test. Remember this one slightly odd reaction when making in-circuit tests: If the transistor-tester says it's good, it probably *is;* if it says it's bad, in-circuit, it may or *may not* be. For a conclusive test, take the transistor out and recheck it. You're going to have to take it out anyhow, if it is bad. There may have been something shunted across the transistor while in circuit.

The transistor-tester should be able to read the beta or H_{fe} of both types of transistors. This is roughly equivalent to the "mu" of a tube. Only on very rare occasions will you find a transistor with a beta lower than its rating. The beta reading is useful mainly for selecting replacement transistors for use in a given circuit. In other words, if a transistor shows a beta reading, it is capable of amplifying a signal. The typical transistor tester will have special dial settings for all types of transistors, from the very tiny types to the big power transistors used in audio.

The second test, and a very important one, is for *leakage.* From experience, I would say that this is the most common and hardest-to-find fault in all types of transistors. Most testers will read what is called I_{cbo}, or collector-to-base leakage. There are a lot of other types of leakage, as you'll see if you read a transistor specification sheets. Some transistor testers will read many types of leakage. However, I_{cbo} is the most common fault, apparently, and the one that can cause the most trouble. This is defined as the leakage current through the collector-base junction when the collector is reverse-biased, as it is in normal operation. This, of course, upsets the base bias. In other words, because of this leakage current the base "loses control" of the collector current; this is guaranteed to produce all kinds of ill effects on operation.

Be aware of the type of transistor! In germanium power transistors, for example, a fair amount of leakage is permissible, sometimes as high as 200 to 300 microamperes. However, in silicon power transistors, practically zero leakage is normal. In some cases, even as little as 10 microamperes of leakage has caused a great deal of trouble. So, know what type you have, before you reject it for leakage.

All leakage tests, with no exceptions, must be made out of circuit. Since we are dealing with leakage currents in the order of microamperes, we can't have anything at all shunted across this junction when this test is made. This includes our moist fingertips, too, by the way. Clip the test leads to the transistor, and place it on the bench, or hold it by the insulation of the leads; keep your fingers off any exposed parts of the test circuit.

Extras

A good capacitor-tester will be very useful. You'll need to find two things about capacitors: their capacitance (which can change value by opening) and their leakage. This is very important, especially in the many cases where you'll find electrolytics used as coupling capacitors. The capacitor tester will also confirm a diagnosis of high power-factor in filter capacitors. However, the scope is a very good "capacitor-checker" for filter capacitors, as you'll soon see.

The Wattmeter

A standard wattmeter, say one which will read up to 300 watts, is a very valuable tool. When you start in servicing any kind of transistorized equipment, plug it into this wattmeter outlet. Then,

Courtesy Sencore, Inc.

Fig. 4-1. Wattmeter that can be used to measure input-current drain.

you will be able to read the actual power being drawn. This will give you a lot of useful information in a very short time. There will be more details later on; Fig. 4-1 shows one.

The Variac

This is simply a variable transformer. It will vary the ac line voltage from zero up to a point slightly above the full applied ac voltage. This can be used to make one special test that can be a big money-saver. Even with battery-powered equipment, it, as well as the wattmeter, can be used. Just plug the Variac into the wattmeter outlet, and then plug your ac-powered test power supply into the Variac. The bench power supply can then be adjusted to deliver any dc voltage needed for certain types of equipment. Fig. 4-2 shows a Variac mounted in a utility box, with an ac voltmeter and outlet.

Fig. 4-2. Variac mounted with ac voltmeter and outlet.

Hand Tools

Your regular hand tools will be fine. In several cases, you'll find that a miniature pair of long-nose pliers will be very useful for getting parts out of, and back into, the numerous tight places you're going to run into in transistorized equipment. An old pair of surgeon's hemostats are very handy; they can be locked onto small parts, and will hold them firmly until released.

When choosing soldering irons, a very small one (30 watt, etc.) is almost a necessity. Get one with a very sharply pointed, conical tip. This kind of tip will get into the tight places you'll find, and make solder joints with much less danger of making solder-bridges between nearby conductors on the PC board.

A desoldering iron, sometimes called "solder-sucker" is another almost essential tool. These are especially useful in removing and replacing small transistors. A hollow tip is placed over the soldered joint, and a rubber bulb on the handle is squeezed, then released. The molten solder is sucked out of the eyelet on the board, leaving it clean for reinsertion of the transistor. It's almost impossible, in some cases, to get all of the leads of a transistor melted and loose at the same time without one of these useful gadgets.

Fig. 4-3 shows the whole group. With these, you can test and repair any transistorized audio amplifier, from the very smallest up to the mammoth types used for musical-instrument amplification.

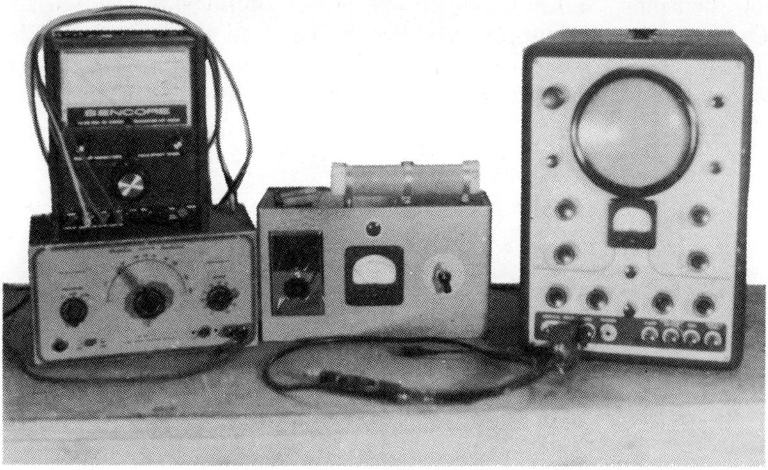

Fig. 4-3. Group of instruments useful in transistor amplifier servicing.

"THE METHOD": TEST TECHNIQUES

The most important "test instrument" on your bench is that mass between your ears. Electronics maintenance work of any kind is 95 percent diagnosis and 5 percent "execution." It requires the ability to *think logically*. No test instrument can make the diagnosis for you; *you* have to do this from the instrument readings. You need two things: one, a good knowledge of the circuitry you're working on, and two, the ability to make tests at key points, getting some information about how the circuit is reacting. Based on this information, you then make the diagnosis as to the most probable cause of the trouble. This is the hard part. The actual repair—the soldering-in of a new part—can be done by a 10-year-old boy.

Since in all kinds of electronics work we always have multiple causes for a single symptom, we must use a logical step-by-step process of elimination. When we start, we have a certain group of parts that could cause the trouble we see. We must patiently and logically eliminate them one by one, and when we do we find the trouble in the shortest possible time. So, the development and use of a logical test method is an absolute essential. We'll give you the basics of our method, as worked out over a period of many years, and you can modify it to suit yourself. Every technician works just a little bit differently, and each should use the way that works best for himself.

Here are the tests, in numerical order, and the reasons for using them. You must have a logical reason for making *any* test. If you don't, you're not "testing"; you're just fooling around!

Incidentally, here's a statistical method, based on many experiments. The stages and circuits most likely to fail, in order of frequency of occurrence, are:

1. The dc power supply
2. The power output stage
3. Drivers and preamplifier stages

If you can make your tests in this basic order, you stand a much better chance of finding the trouble faster. Now for the detailed method.

"WHAT'S THE MATTER WITH IT?"

Find out *what* the trouble is. In other words, why was it brought in for repair? You get this information by asking the customer. Don't use those exact words; if you do, he'll say, "That's what I brought it in here to find out!" Try asking "What's it doing?" Even if the reply is "Nothing!" that's some data; it is completely dead. In any case, you will find out *what* it is that he wants fixed. No output at all, distortion, and, most important of all, whether the thing is *intermittent*.

DETAILED TESTS

The Logical Method

First, hook it up and try it, to verify the customer's complaint. If it's dead, plug it into the wattmeter and turn it on. If you see no reading at all, there's a blown fuse or something like that. If the wattmeter needle hits the peg, there's a bad short.

If it's drawing no power, check the logical things: first, ac power input, open line cord, bad switch, blown fuse, etc. Ohmmeter checks will confirm your first diagnosis.

If it shows a short or overload, check the dc power supply with the ohmmeter for shorted rectifier diode, shorted filter capacitor, shorted output transistor, bypass capacitor, etc. Make sure that the dc power supply is working. If it isn't, none of the other circuits *can* work.

If you have normal dc power supply but no sound at all in the speaker, you've eliminated the first possibility in our statistical list. At this point, the trouble *can* be anywhere in the circuit, but the output stage is the most likely, since it's the one handling the greatest amounts of power; its parts are under greater stress, hence more apt to fail. So, check it first.

Dc voltage readings here will give you a good idea of what's going on. One key test for output-transformerless circuits is the dc voltage at the midpoint, where the speaker is connected (also one of the easiest to locate). Any dc voltage error indicates trouble in the output transistors or output stage. Correct dc voltage at this point shows that the output transistors have a very good chance of being in good working order.

If the output stage is probably working, but there is still no output, then the problem must be somewhere else in the circuit. (Don't forget to check for an open speaker before you leave.) It's easy to narrow this down; it's somewhere between the output and the input. That sounds a little ridiculous, but is demonstrably true. So, how do we find out *where*? Use the logical method again.

At this point, we have two facts: our dc power supply is working, and the output stage is working. We've eliminated two of our three probabilities; the list is narrowing down. Now, we remember another fact; the "signal path" between the input jack and the speaker is a SERIES CIRCUIT. Somewhere along the line, this circuit is open. The signal can't get through, so we make logical tests to find where it's being blocked.

(If this sounds a little bit simple to you, that's because it is. All of these tests and measurements are aimed at finding ONE simple little thing, a bad transistor, an open resistor or capacitor, etc. That's what is causing the trouble.)

For the next step, feed a test signal into the input. Check its waveform and amplitude with the scope. At this time, we're not interested in anything but the signal *amplitude*. We're trying to find out where the signal *stops*. Note the amplitude at the input; set the scope's vertical gain to make a pattern say an inch high. Now, move on to the input of the first amplifier stage, the base of the preamplifier transistor. If the signal is there, go on to the collector. If it's there, go on to the base of the next transistor.

Right now, we're not really interested in checking for *gain* through these stages. What we're looking for is a stage that has input signal but no output—in other words, a *dead stage*. That's where the signal stops, and *we* start. With the right service data for this amplifier, within about *30 seconds* you should be able to locate and check each amplifier stage and find out which one is not passing the signal.

Once you arrive at this point, lay down the scope probe and pick up the dc voltmeter probe. Check for the presence or absence of the normal dc voltages. Incidentally, this should be done with the audio signal turned off. Practically all schematics give you the *no-signal* dc voltages on the transistors. While there isn't the amount of voltage variation between no-signal and signal conditions that you find in certain tv circuits, there may be some. So, just to make sure, turn the signal generator off.

Statistically, the chances are that your dc voltage readings will not be just a little off; if they're off enough to kill the stage completely, they'll be off far enough that you'll be able to tell right away. For example, a collector that should have +25 volts on it will be zero, and so on. At this point the power can be turned off, and you can check the most likely suspect, the transistor. Regardless of all the myths that we have heard about transistors being "immortal," they aren't. We have found that the majority of defects in transistor circuitry are due to defective transistors! An in-circuit test here will tell you whether the transistor itself is bad.

Signal Tracing

What you've been doing is signal tracing, and it's probably the quickest way of locating the cause of the trouble. Practice this by checking out an amplifier known to be in good shape. For this type of test, you don't even have to readjust the vertical gain of the scope. Set it for a small pattern on the input, and then let the pattern go off the screen if it wants to. If it does, you know that you have considerably *more* signal at that point than you had at the input, and that is ALL you want to know right now. What you're looking for is a stage with signal on the input and a nice straight line on the output. In other words, no signal at all. You can run the gain down if you want to, but you don't have to. Later on, when we're checking for gain, you can make more accurate tests. At this time, all we're interested in is the answer to the vital question "Is it there or isn't it?" and as you can see, the scope will really give you the answer in a hurry. In just a minute, we'll go through a full signal-tracing procedure all the way through a typical amplifier, and see how you can get the answers to other questions just as easily.

FAST FAULT-FINDING WITH SIGNAL-TRACING TEST METHODS

The preceding signal-tracing test was a bang-bang type, designed only to find the location of a faulty stage in the shortest possible time. The same technique can be used, a little bit slower and more

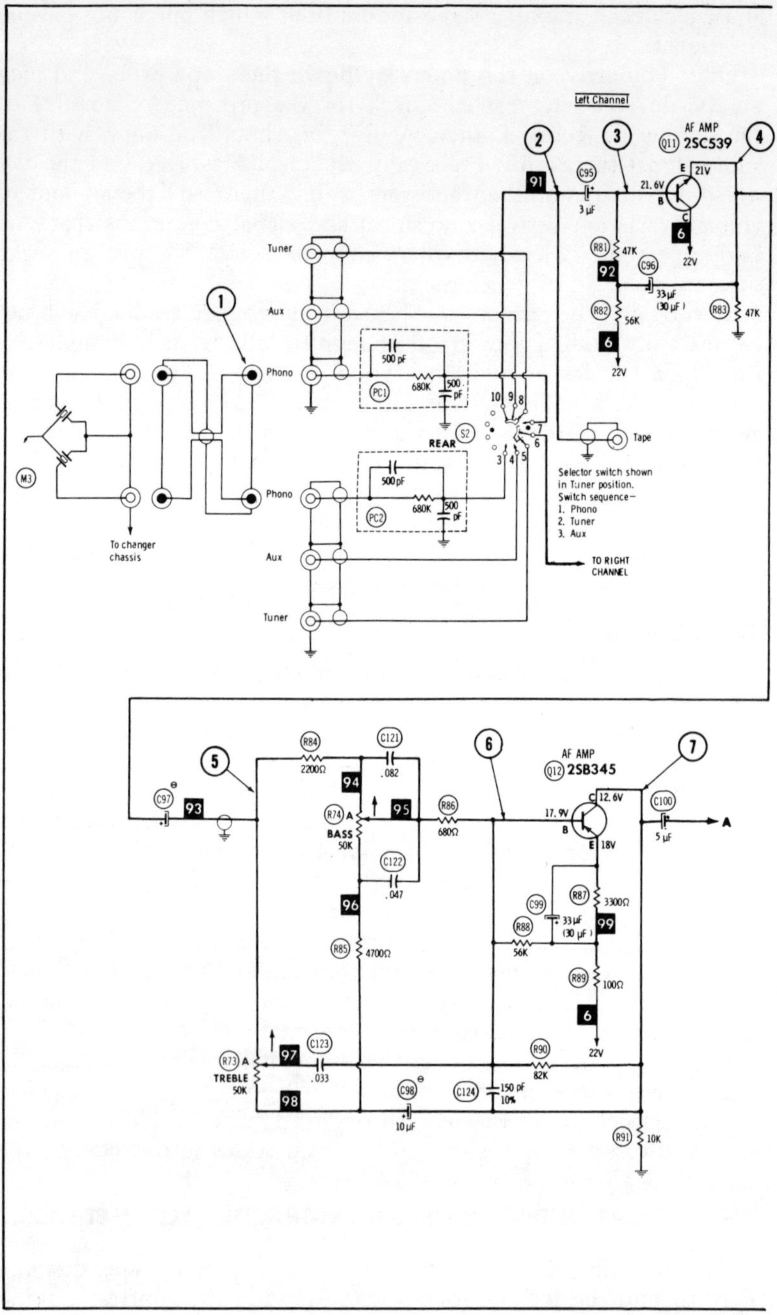

Fig. 4-4. Stereo

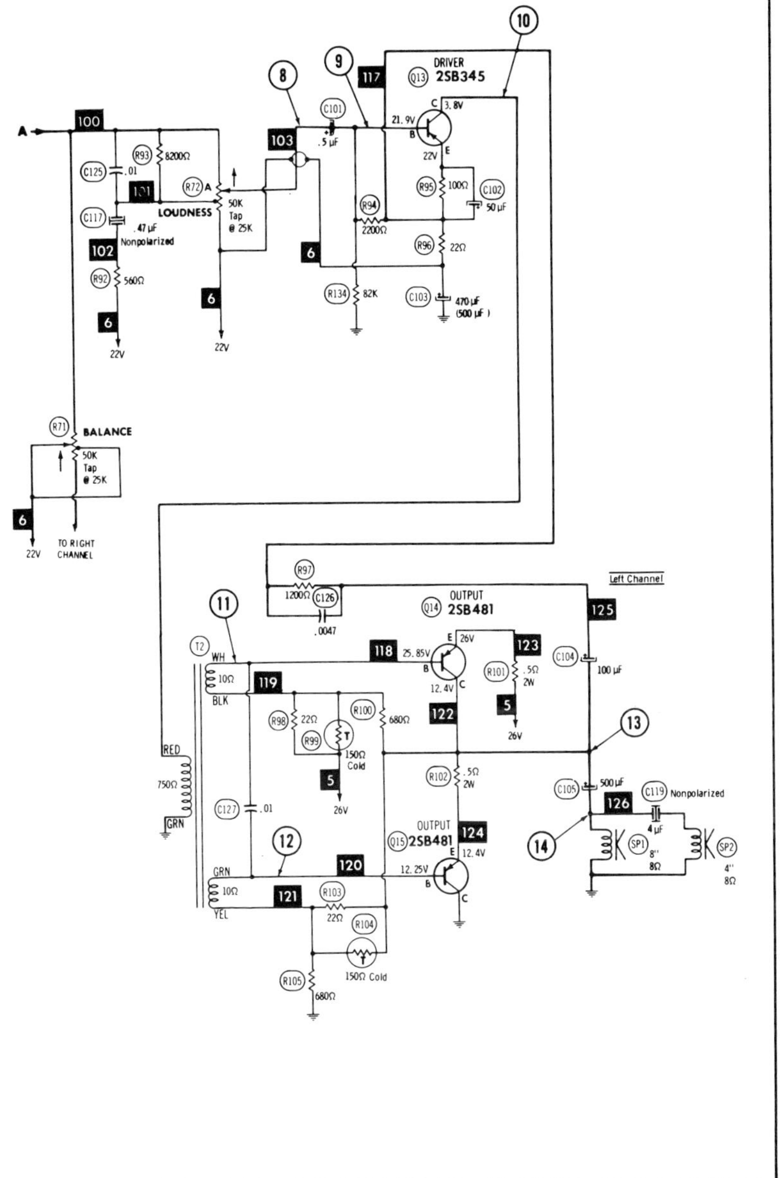

amplifier circuit.

detailed, to check every part in the signal path of any kind of amplifier from a 1-watt mini to a 300-watt maxi. All amplifiers are exactly the same as far as the signal is concerned. Power output, voltage, polarity, types of transistors, circuits, or anything else—makes absolutely no difference. They are all precisely the same when it comes to the *signal*. The bigger amplifiers just have more stages of amplification, that's all.

Let's follow this signal path through an actual amplifier and see how it works. Fig. 4-4 shows the amplifier circuit. You can see the circled test points. Here is a list of each of these points, and the reasons for making the test at that point. Follow it through, one step at a time:

1. Establish the test-signal level on the scope, so that you'll know what you're looking for. Set the pattern here for any desired height; say 1 inch. Actual signal should be about 50 millivolts.
2. The same signal-level as the input, at this point, shows that the signal is going through the selector switch, or any other coupling devices.
3. Same signal as input. Shows that coupling capacitor C95 is not open. This is the base input signal for af amplifier transistor Q11.
4. This is the output signal from the first amplifier stage. Should be about the same amplitude as the input signal, minimum. (Note that *this* stage is bootstrapped, and is an emitter follower, to raise the input impedance. So, there will be hardly any signal *voltage* gain. If you break even, get out as much as you feed in, fine.)
5. Checks for signal getting through second coupling capacitor C97. Same amplitude as at Point 4.
6. Checks for signal getting through tone controls to base of second af amplifier stage, Q12. No loss in amplitude, or not much.
7. Checks for signal amplification of second af amplifier transistor. This is a common-emitter stage, so we should find some voltage gain. Normal is from 3 to 5 times or more, depending on type of transistor, applied voltage, etc. Reset scope vertical gain to show 1-inch height pattern.
8. Checks for signal getting through coupling capacitor C100 and loudness control. Check setting of loudness control. Should show no loss of signal amplitude.
9. Checks for signal getting through coupling capacitor C101. This is the input (base) of the driver stage, Q13.
10. Checks the signal level on collector of the driver transistor. The absolute voltage gain will vary, depending on the

impedance of the load, which is the primary of the driver transformer. If this is low, and the collector current is high, you may find a small voltage loss. Ordinarily, if you break even, it's all right.

11. Checks for signal voltage on base of top output transistor. Also check condition of driver transformer primary and secondary. May show small voltage gain, may not. This depends on design.
12. Checks for drive signal on base of lower output transistor. Also checks lower secondary of driver transformer.
13. Checks for action of output stage. This signal should be quite a bit higher in amplitude than base signal. Also checks for possible trouble in either output transistor. This signal should be a clean sine wave, an exact duplicate of input waveform.
14. Checks for signal through output coupling capacitor. (Note: if this capacitor should be open, you won't get any signal at this point. If the speaker is open, you probably won't get signal here either; the capacitor won't be able to charge and discharge, so the output stage won't work. The signal will probably be missing in Test 13, also. Dc voltages will be badly upset.

There you are. This whole series of tests takes a lot less time to do than it does to read about it. After a little practice, you will be able to go through the whole thing in a very short time and land on the cause of the trouble. The same test sequence can be used to locate loss of gain, or distortion. Like all other detailed test-series methods, there are many possible shortcuts and quick checks. Let's take a look at some of these and save even more time.

SHORTCUTS AND QUICK CHECKS

You don't have to run the whole series of tests just given, in every amplifier. By making close observations of exactly *what* a given amplifier is doing or not doing, you can jump right into the middle of the test sequence, and skip many steps. This will pin the trouble down a lot quicker.

For example, if you hear no sound at all in the amplifier of Fig. 4-4, with a test signal fed to the input, turn it off and disconnect the speaker. Check the voice coil for continuity with an ohmmeter. If you get the right reading, or hear a "tick" sound when you touch the ohmmeter leads to the terminals, the speaker is capable of making noise. The amplifier *is* dead. You've eliminated the speaker.

Now, take a scope check right in the middle of the circuit, say at Point 8 or 9. If you get a fairly good sized signal here, you've checked the whole front end of the amplifier at one swoop. Two transistors, three coupling capacitors, and many other parts: they're all working. Now, quickly check from the driver on through the output stage and you'll find out where the signal is being lost. Fig. 4-5 shows a detailed view of this much of the circuit.

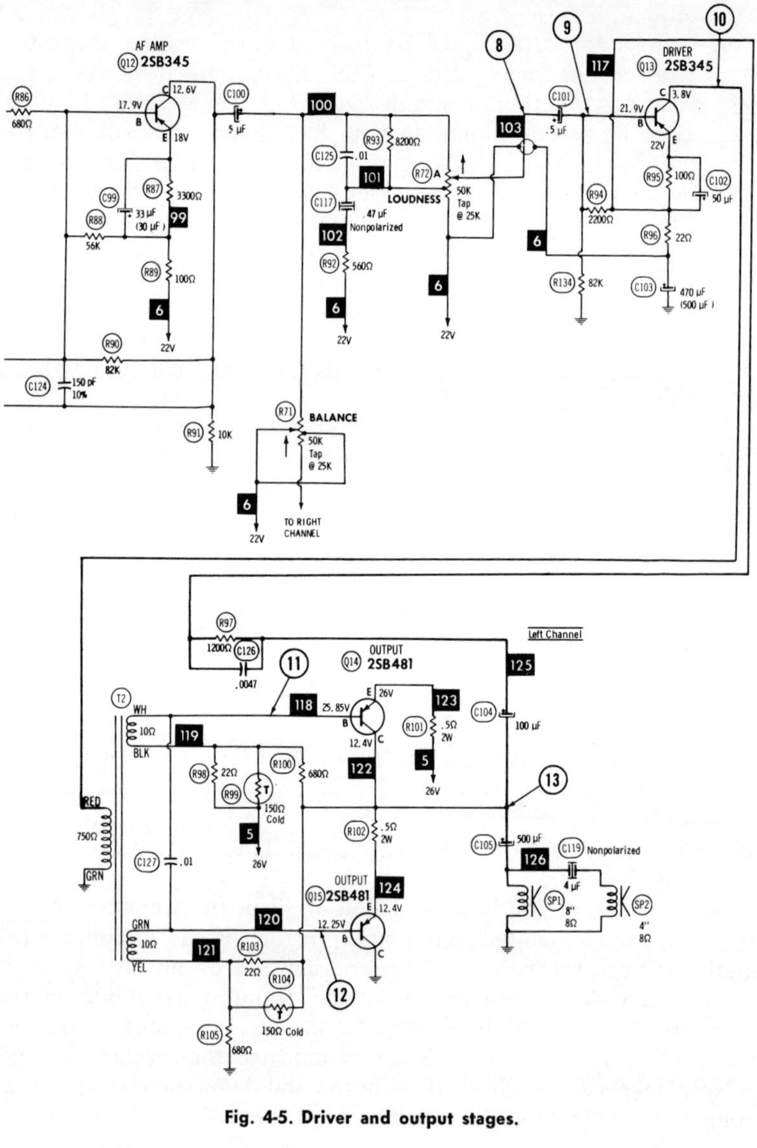

Fig. 4-5. Driver and output stages.

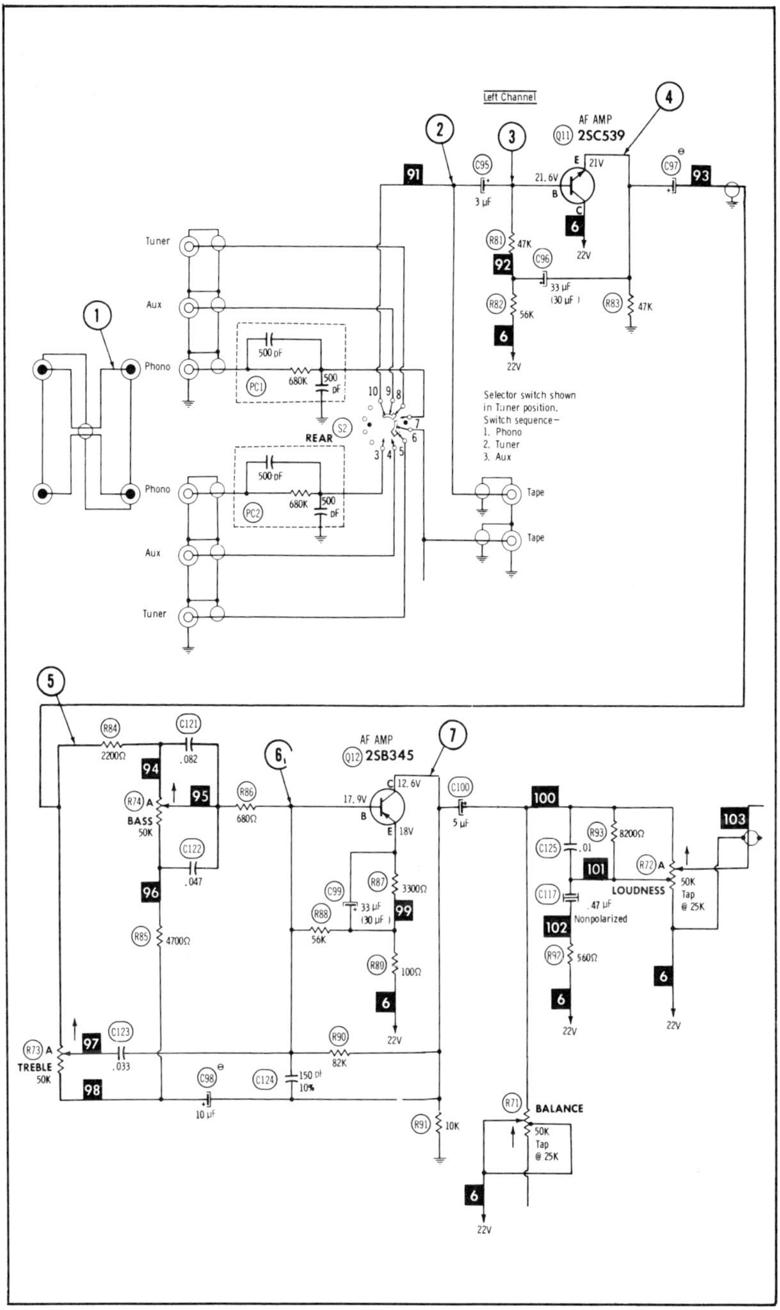

Fig. 4-6. Working area for "front end" tests.

If you get no signal at Point 8, turn around, and go back toward the input. At this moment, you don't know whether the output stage is working or not, but you'll never find out until you get the front end cleared up. Fig. 4-6 shows the "working area" after this test. Check these points in reverse order until you find a signal. Then, stop and check the last stage or part that you passed—that's it.

There comes a time in the signal test sequence, when you lay down the "signal indicator," the scope, and start making a voltage analysis. You've discovered the point where the signal is lost, and now you must find out why it is being lost. So, we pick up the DC VOLTS probe of the vtvm. NOW we have to take voltages, polarities, etc. into account, for they can be vastly different, depending on the power output, type, and circuit used in this amplifier. However, the basic *symptoms* will still be the same, especially in OTL stages.

Let's take a real case history. Signal tests showed ample drive signal on the bases of the output transistors but no signal at all on the output. The only things between these two points being the output transistors themselves, the dc voltages on them were checked. A quick check of the supply voltage in this circuit, on the emitter of the top transistor (Q14, Fig. 4-5), showed this to be a normal +26 volts. That's good.

Second check, on the midpoint of the two transistors, showed practically no voltage at all here. Trouble located. Now, why? There are, as usual, several possibilities, but the most likely one is an open top transistor. Checking the emitter-collector junction in both directions with an ohmmeter showed no difference in the reading, which was a very high resistance. A normal transistor will show a low reading one way and a high reading the other—this is the "diode effect." Remember it. We'll use it often as we go along. Pulling the top transistor and checking it with the transistor tester verified the diagnosis.

Both transistors in this channel were replaced with a matched-pair of replacement types. These are available in several different brands of replacement transistors. In this case, Sylvania ECG-131MP (Matched-Pair) were used. You should always replace both transistors when either one goes out: the replacement might not match the characteristics of the remaining original transistor.

Let's pause for a moment and see how this test sequence and method can be applied to all stages of this basic type. For example, in this circuit, the dc voltage is fed to the emitter of the top transistor, its collector is tied to the emitter of the bottom one, and the remaining collector is grounded. Both are pnp, so this is a "stacked" circuit. If the transistor connections were reversed, with the dc voltage fed to the top *collector*, and the bottom *emitter* grounded, the circuit would be exactly the same. Since these are pnp transistors, the supply voltage would then be fed to the collector as a negative voltage, but every circuit reaction would be the same.

If the same circuit were used with npn transistors, once again everything would be the same. The voltage polarity would change in this case so that the collector would be reverse-biased, but otherwise the same as far as circuit reactions, voltage tests, etc. Even in the complementary-symmetry type of OTL circuit, where the two transistors are of opposite types, one pnp and the other npn, you will still get the same fault-reaction. If there's no dc voltage on the midpoint, the top transistor is open.

For some reason unknown at present, the top transistor in these circuits seems to be the one causing the most trouble. Possibly due to the fact that it is the one which gets the initial voltage pulse from the power supply, and thus is more apt to break down, blow the junction, or short. At any rate, in actual replacement work, this one provides most of the trouble. The lower transistor may be leaky, etc., but if one of them is blown out, it will be the top one.

Quick-Checking Through the Other Stages

Now, we've got the output stage fixed. Let's see if there are quick checks that can be used on the remaining stages. Our signal-tracing tests showed that there was no signal on the bases of the output transistors, but there was ample signal on the driver base. So, we've pinned the fault down to a single stage once again.

Fig. 4-5 shows the driver stage circuitry. If we have signal on Point 9, but none on Point 10, we know that the driver transistor is not conducting. Note that I did not say that the driver transistor was bad. We don't know that—not yet. We do know that it is not amplifying the signal, therefore it isn't conducting current as it should. Now, we go to dc voltage tests.

The collector of the driver goes to ground, through the primary winding of the driver transformer. However, this is not a common-collector circuit, but a common-emitter circuit; the signal is taken off at the collector. The dc voltage on the collector is developed across the resistance of the driver transformer primary. No collector current, no dc voltage. Our voltmeter verifies this; zero volts. So, the transistor definitely isn't conducting. (Don't be afraid to duplicate the tests, as we just did. While they show the same thing, it never hurts to have confirmation.) Incidentally, this test also tells us that the transformer primary winding is not open. If it were, we would have read a small dc voltage on the collector, since the other elements of the transistor have their normal voltages.

We have now eliminated things until we have one of two possible conditions. One, the transistor is open; two, the transistor is good, but biased to cutoff. We can check this in any one of several ways.

Probably the quickest way would be to turn the power off and make an in-circuit check of the transistor. Or, while we have the DC VOLTS probe in our hand, check the voltages on base and emitter. If they read exactly the same, say +22 volts each, this means that there is zero volts from base to emitter and the transistor is cut off.

This voltage is "clamped" by a voltage-divider network from the dc voltage supply on the emitter. So, if the transistor *is* open, the base bias will still be the same, just a little bit negative with respect to the emitter. (This figure shows us one more thing, it is about 0.2 volt, so the replacement transistor must be a germanium type. If this had been about 0.6 volts, it should be a silicon.)

You could also check for a diode effect on the transistor collector-emitter junction. However, with only the 750-ohm resistance of the winding and the low resistances in the emitter circuit to the dc supply, this can be confusing. The dc power supply will normally show a pretty low resistance to ground, due to conduction through the rectifier diodes. This is especially true in circuits like this with a bridge rectifier. No matter which polarity your ohmmeter battery happens to be, one of the diodes in the bridge will be forward-biased and show a very low reading. Therefore, the transistor tester will be the fastest and easiest way to make this test. Incidentally, it's a very good idea to recheck the original transistor after you get it out of circuit. Never be afraid to check your own ideas. You might be wrong. If so, better find it out early, and save trouble!

This type of testing will work in all of the stages from this point all the way to the input. In the preceding stage, Q12, note that the collector is grounded through a 10,000-ohm resistor. Aside from the numerical difference between this and the 750-ohm transformer winding, there's no difference at all in the test methods and results. (You can get a diode-effect ohmmeter test on this collector, due to the greater resistance.)

In other types of common-emitter stages, such as the one shown in Fig. 4-7, we still have an "indicator" of current but it's in a different place. This figure shows a common-emitter stage with the dc voltage fed to the collector. The emitter goes to ground, through a 330-ohm resistor, bypassed with a 50-μF electrolytic capacitor. Notice the dc voltage on the emitter? While this is only −1.35 volts, it is developed by the collector-emitter current, again. So, it serves our purpose just as well. If there is *no* collector current, the emitter voltage will be zero, and we have our "indicator."

SHORTED TRANSISTORS

We've been talking about open power transistors and other types. Now and then, you will find transistors with collector-to-emitter

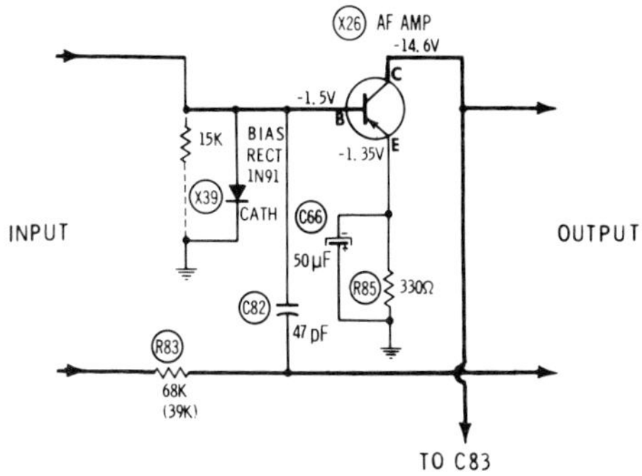

Fig. 4-7. Common-emitter circuit.

shorts. The signal symptoms will be exactly the same, but you'll get a different voltage indication. You'll find the same dc voltage on collector and emitter. If there are small emitter resistors in the circuit, these will probably be burned up. This applies mainly to power transistors. Driver and preamplifier transistors will not pass enough current, even under short-circuit conditions, to do much damage to the resistors.

It is possible to have a short between any of the three elements, or all at once, of course. The same dc voltage on base and collector indicates a base-to-collector short, and so on. However, emitter-to-collector shorts seem to be the most frequent. The ohmmeter is a good tool for finding shorted transistors. You will normally get a very low resistance reading both ways, especially across silicon transistors—almost zero ohms.

A little later on, we will run across some unusual symptoms due to shorted transistors and circuit peculiarities. However, in the typical case, there will be no signal getting through a shorted transistor. The key test is the presence of the same dc voltage, usually the collector voltage, on one of the other elements, base or emitter.

THE "WEAK" AMPLIFIER

The preceding tests were aimed at finding where the signal stopped in a completely dead amplifier. You can use exactly the same test methods and instruments if the complaint is not a dead amplifier but one with low gain. This isn't as common in transistor amplifiers as it is in tube types. Transistors do not "get weak" (i.e., low emission) as tubes do. You will seldom find the "clear but weak" symptom so common in tube amplifiers (in the majority of cases, a loss of gain will be accompanied by fairly severe distortion). However, it's not impossible. Certain things can happen which will give this kind of effect, as you'll see as we go along.

Low Voltage

One thing that can cause low output, with fairly good tone, is low dc supply voltage. For example, if the input filter capacitor should open, the dc voltage would drop to half of its normal value or less. This will reduce the power output, but in most cases won't cause too much distortion. This is very easy to find if you use the proper test sequence and check the supply voltage first. Oddly enough, an open *input* filter will not increase the hum level as much as you would expect if the output filter capacitor is still good.

Bias Upsets

Low volume can also be due to bias upsets on most stages. This is sometimes due to resistors changing in value, because of heat, etc. This isn't a very common type of trouble. Most of the resistors used in transistor amplifiers will be low in value as compared to tube-type circuitry. The higher the value of the resistor, the greater its tendency to drift.

Shorts in high-current transistors will generally damage the resistors in the circuit, especially the emitter resistors. However, most of these will go entirely open and be easy to find. Some are of the "fusible" type, designed to burn out on overloads. Carbon resistors which have been burned are easy to spot on a visual check, but the low-wattage wirewounds won't discolor; you'll have to check these with the ohmmeter.

Dc voltage readings around the suspected stage will help. Look for base-emitter voltages that are slightly below the "conduction bias" for the type of transistor being used. For example, in a silicon transistor with only about 0.3 volt bias, 0.6 volt should be the minimum bias, even for Class-B operation. The scope will show up the resultant distortion and the dc voltmeter will catch the incorrect bias.

Leaky Transistors

Probably the most frequent cause of low volume is transistor leakage. Collector-base leakage seems to be the most common. This will always cause a severe upset of the bias, and heavy distortion because of clipping or saturation.

You cannot check transistor leakage in-circuit. There are always so many shunt paths that you'll read a large "leakage," but this is a "fake" symptom. To make a valid leakage test, the transistor must be taken out of the circuit and checked with a good transistor tester. Germanium transistors will always show a small leakage, somewhere around 10 microamps or less in the small-signal types (drivers, preamps, etc.) and up to a couple hundred microamps in the big power transistors. Silicon transistors of any size and type should show zero leakage on the average transistor tester. Even as little as 10 microamperes of leakage is enough to upset the operation of a power transistor. A loss of base control in the transistor and the resulting bias upset will cause heavy distortion. Read the dc voltages, replace the transistor and read them again, and you'll be sure.

Low Output Plus Distortion

In conventional common-emitter stages one frequent cause of low output with high distortion is an open emitter-bypass capacitor. In the circuit of Fig. 4-8, note the large bypass capacitor across the emitter resistor. This will run from 50 μF on up to 100 μF or more. Naturally, with this big a capacitor directly to ground, there should be practically no *signal* at all on the emitter. The bypass capacitor is used to preventing any shifting of the operating point because of signal voltage variations on the emitter.

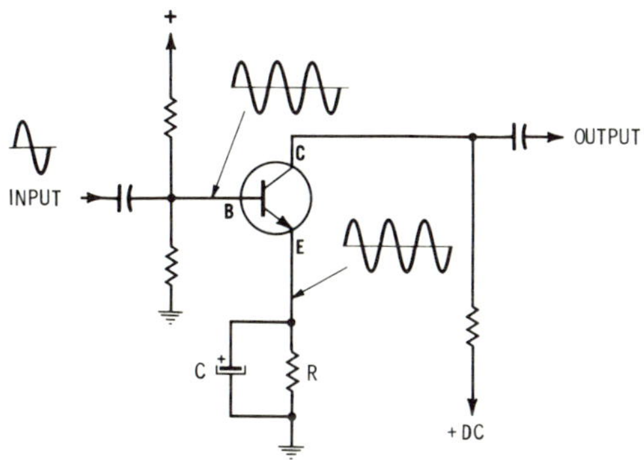

Fig. 4-8. Test for open emitter-bypass capacitor.

The quickest test for this is a scope check on the base and emitter. If you read the *same* value of signal on both points, the emitter-bypass capacitor is open. This will not have too much effect on the dc voltages. It will cause a very high degenerative feedback, resulting in high distortion. The distortion is not due to the degeneration in *this* stage but to the severe "under-drive" condition in following stages. This throws these transistors onto the wrong parts of their curves, and this causes the distortion.

Clipping

Transistors are very easy to drive into clipping due to the their very sharply curved characteristics. Voltages are comparatively low, and even a little bit too much input signal will drive a transistor into collector saturation. This causes clipping of the tops of the signal waveforms. A small variation of the bias will let the transistor "bottom out" and will do the same thing for the other peaks.

The scope and a test signal are the fastest means of finding the cause of clipping. Feed in a good sine-wave test signal of any easily readable frequency and simply follow it through the amplifier stages. When you get to the stage that's causing the trouble, you'll know it. The signal will be very definitely "flat-topped" from that point on through the rest of the amplifier. Fig. 4-9 shows the general appearance of a clipped signal.

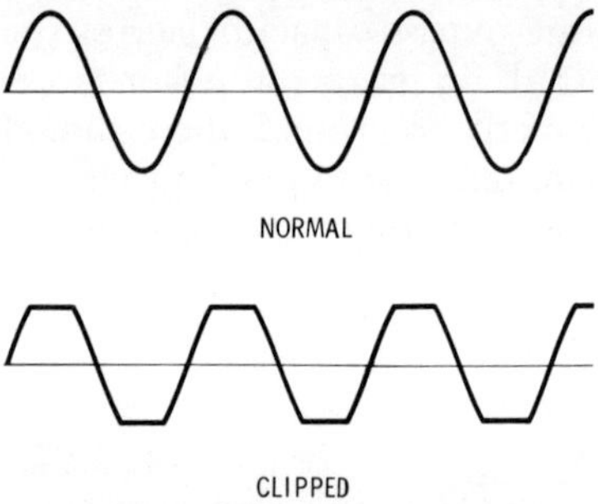

Fig. 4-9. Scope pattern of clipped signal.

CAUTION: You can cause clipping of a signal, by using too high a level on your test signal. The normal inputs from such things as microphones, guitar pickups, and phono cartridges are very small. Mikes and pickups will be down in the millivolt range. Some high-output phono cartridges, especially the ceramics, may have an output up to 1.0 volt or more. So, the first two inputs are fed into high-gain input jacks, and the phono cartridge into one having reduced gain, just to avoid the chance of overload and clipping.

To be certain of this, never use an input greater than 50 millivolts. You can make a rough check on this by simply feeding the signal into the input of the amplifier, and checking the output of the first preamplifier stage for clipping. Reduce the signal level with the attenuator until there is no flat-topping. If the signal is getting through the first stage without distortion, it will probably be all right.

Many audio signal generators have trouble reducing their output that low. If you find this is the case, you can make up a voltage divider with a 950- and a 50-ohm resistor, as shown in Fig. 4-10. Set the signal level across the divider at 1.0 volt; your output is taken from across the 50-ohm resistor, and is exactly 50 millivolts. This is as close to a standard test signal as we have, at this time. No one can agree on exactly what the specifications of a standard test signal should be, but quite a few manufacturers use the 50-millivolt level for all of their testing.

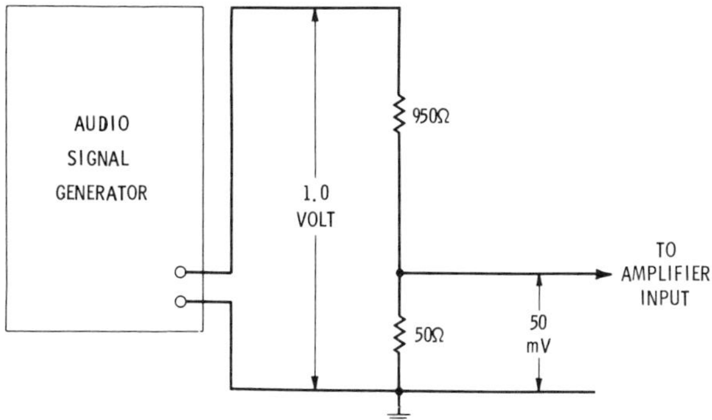

Fig. 4-10. Audio signal voltage divider.

Tone Controls

When making distortion tests, be sure to set the tone controls, if any, in their FLAT position. Excessive bass or treble boost actually distorts the signal. You can even come up with something like the waveform of Fig. 4-11, simply by adjusting the tone controls incorrectly! The input was a good sine wave. This is especially bothersome when you're testing with square waves. Feed in a square-wave signal, and move the tone controls, and you'll see what we mean very quickly.

FILTER CAPACITOR TESTING WITH THE SCOPE

One part that will give you quite a bit of trouble is the electrolytic filter capacitor. If one is shorted, of course, it won't be too hard to find. Just follow the trail of burned resistors, and the shorted capacitor will be right after the last one. However, open capacitors, or capacitors with a high power factor, can cause some weird troubles and be much more difficult to locate, unless you use the right test equipment.

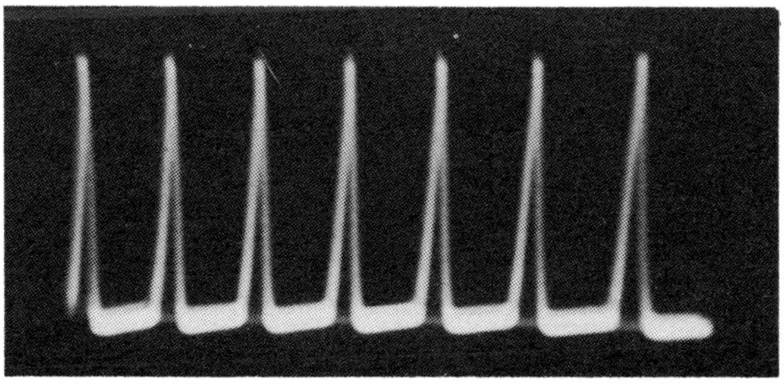

Fig. 4-11. Distortion due to incorrect setting of tone controls.

A capacitor having a high power factor is one which has developed a little too much leakage. This reduces its filtering efficiency and allows high levels of feedback through the dc power supply. Since the dc power supply lines are common to all stages, they must be kept at absolutely "zero impedance" to ground. If there is any impedance present, the signal-return currents will develop *voltages* across it; these undesired voltages will then be *fed back* into other stages where they'll arrive in the right phase to cause oscillation, distortion, and a multitude of other problems. Fig. 4-12 illustrates this.

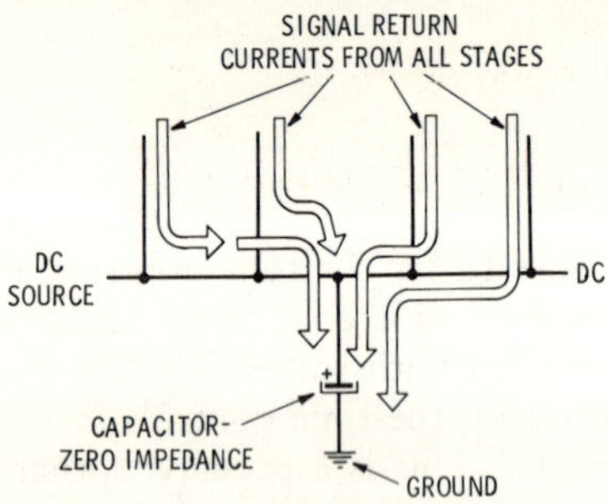

Fig. 4-12. Electrolytic capacitor as a common return path for signals.

A good electrolytic capacitor, especially with the high values used in transistorized power supplies, has a very small "ac resistance" (impedance). Putting several of these in parallel reduces the impedance even more. However, if these capacitors develop high power factor, their impedance rises quite a bit, and we get into trouble.

There is a test instrument that will reveal the presence of any unwanted signals on the dc supply line, and thus check its filtering efficiency. This is the scope. The test is ridiculously simple; the filter capacitors are used to prevent the appearance of undesired signals on the dc supply lines. Therefore to test them for efficiency, we simply touch them with the scope probe and look for any sign of *anything* in the way of signals; if you see anything at all, it's wrong! With vertical gain turned to maximum, the normal pattern, indicating that the filter capacitors *are* good, is a simple straight "dc line" on the scope. Fig. 4-13 shows what you'll see if there is a bad filter capacitor; this pattern had a peak-to-peak amplitude of over 100 volts. Note the presence of a very high frequency oscillation in between the 120-Hz peaks of the ripple; this was caused by one of the stages going into violent oscillation. Incidentally, the dc voltage on this line was normally only 50 volts; the higher voltage was developed by the oscillating stage.

Even if the amplifier isn't in oscillation, you can catch weak capacitors that cause hum or ripple in the output. The pattern then

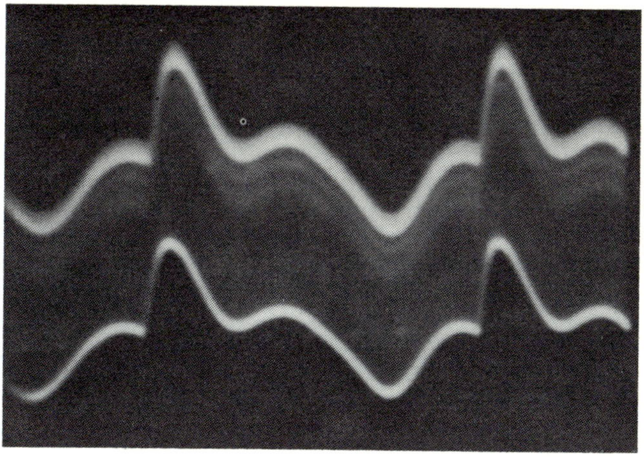

Fig. 4-13. Pattern of oscillation caused by open filter capacitor.

will show you a basic 60- or 120-Hz frequency. It will sometimes be almost a sine wave, but more often will look something like a sawtooth. Ripple levels in well-filtered transistor power supplies should be very low. Even 0.5 volt pk-pk is sometimes too much.

To check, connect the scope probe across any of the filter capacitors, except the input capacitor. At this one, you will see a small amount of normal ripple, usually something like 10 volts pk-pk. On the output filter capacitor, this ripple should disappear completely. With the scope clipped to any suspected test point, note the pk-pk value of the ripple. Then, turn the power off, clip another capacitor of the same size, or larger, across this one, and turn the power on. Note the amount of ripple. If the original capacitor was low in capacity or open, you'll get your straight line back.

NOTE: *Never bridge a test capacitor across the dc power supply in transistor equipment with the power on!* The resultant charging surge will damage something. Capacitors are so big that they take a high surge of current when connected across a dc power-supply, and transistors do not like transient spikes or surges in their power supplies. With the instant warmup of transistor equipment, you won't lose any time. When all filter capacitors charge up at the same time, there is no surge.

In certain cases, you'll find that bridging a test capacitor across the circuit will reduce the ripple slightly, but will not bring it back to zero. If you run into this, the original capacitor probably has high power-factor, instead of being open. The only reliable test for this is to disconnect the original capacitor, hook in the substitute, and recheck. If the ripple now is much lower or gone completely, that was it. The high leakage of a high-power-factor capacitor will still develop enough feedback signal to leave some ripple showing, even with a good capacitor across the circuit. To make sure, pull it.

WATTMETER TESTING

The wattmeter on the ac input is probably the most useful single test you can make, and certainly the simplest. Plug it in, turn it on, and look at the wattmeter. Checking the normal power drain as given in the service data will tell you very quickly whether there is trouble in this amplifier, and, with just a little interpretation, what kind.

A normal wattage reading, if the amplifier is completely dead, tells you that there is a break in the signal-path. Something is stopping the signal, but is not affecting the loading on the power-supply. This could be an open coupling capacitor, for example. So, you don't need to waste any time looking for shorted filter capacitors or things like that. Start right in with signal-tracing tests.

A wattage reading of double or more the normal drain shows you that there is a short or a very high leakage somewhere in the circuit. Amplifiers in this condition are generally very dead. Under these circumstances, you wouldn't start in signal tracing; start looking for a dead short somewhere, circuits with no dc voltage on key points, overheated resistors or transistors, etc. Shorted rectifier diodes are very common causes of trouble. Check them with an ohmmeter. The silicon diodes commonly used in the power supply will give you a nice clear indication when shorted; they will read practically zero resistance both ways. When they do short, they don't mess around!

If there is doubt as to the location of the short, a good quick check is to lift the main lead from the dc power supply from the load, and turn the unit on again for a moment. If the wattmeter shows a very small drain now, the dc power supply is all right, and the short is in one of the load circuits. The most likely cause is a shorted power transistor in one of the output stages. Ohmmeter checks here will clear this up in a hurry. Power transistors, like the silicon diodes, normally go to a very low- or zero-resistance short both ways.

A reading of about half normal power shows an open circuit somewhere, probably in a circuit that normally draws a good deal of power—for example, one channel of a stereo amplifier with an open top transistor in the OTL output stage. This can also be due to an open input-filter capacitor which drops the dc voltage to about half its normal value, reducing the power input by a like amount. One quick dc voltage reading on the power supply will prove or disprove this idea.

In circuits with a full-wave bridge rectifier, one open diode will reduce the voltage, but not as much as you would expect. This will show up as a slight drop in input wattage, and there will also be a decided increase in hum. In solid-state rectifier circuits with voltage doublers, an open input capacitor in the half-wave circuit will cut the dc voltage to a very low level. In full-wave doublers, an open shunt rectifier will reduce the voltage to half normal or less, and the power with it. If the series rectifier should open, there will be no dc voltage output at all and practically no power drain. Once again, these conditions can be quickly checked out with one dc voltage reading on the power supply.

The wattmeter is very helpful for checking amplifiers after the power-output transistors have been replaced. If the input wattage is above normal, the biases on the output transistors should be checked; this could indicate that they are drawing too much current and in time will go into thermal runaway and destroy themselves. This type of trouble is often due to a leaky driver transistor in direct-coupled stages, to a burned bias resistor that was overlooked, and that sort of thing. There is a much safer and easier way of checking after power transistors have been replaced.

REDUCED-POWER TESTING AFTER PARTS REPLACEMENT

The Variac, or variable-voltage transformer, is an invaluable tool for making certain tests. One of them we just mentioned, checking out high powered amplifiers after replacing shorted output transistors. You can never be absolutely sure that those big transistors failed all by themselves! There may have been something in there that you didn't find. Since transistors can destroy themselves in such a short time, you'd better be sure before you turn the power on. (Unfortunately, the ohmmeter is of very little help here, due to the very low resistances present in the power supply and other circuits, even when they're normal.)

If there is some trouble still in there, like a bad driver transistor, etc., the new transistors will blow out just as rapidly as the originals did. Average time, maybe 1/50th of a second—far too quickly for circuit breakers or fuses to respond. And, there you sit with four shorted transistors; two belonging to the customer, two to *you*! Believe me, this is a condition to be avoided at all costs.

There's an easy way of making sure. Turn the Variac dial to zero volts (off), and plug the amplifier into its outlet. Plug the Variac line cord into the wattmeter if you want to. As an alternative, open the dc power-supply circuit and connect a dc milliammeter in series with it, or do both. Note the normal resting-current for this circuit as given on the schematic. Don't use test signals yet. We want to find out if the amplifier will take its normal resting current before proceeding.

Assume that this current is 150 mA. Now, if you do *not* have any troubles in the amplifier, you should be able to get the line voltage up to full value without reading *more* current than this, or drawing more than normal wattage. However, if there is a short in there, we do not want to put the full line voltage on it without some kind of protection.

Start turning the Variac dial up slowly. Watch the wattmeter or milliammeter. You can use a dc voltmeter too, if you want to, but the *power* consumption will tell you much more. With solid-state diodes in the power supply, the dc power will rise slowly and almost linearly with a rise in the applied ac voltage; you don't have to wait for a cathode to warm up.

Raise the line voltage slowly. If you get up to say about 50 volts, and the circuit is drawing almost normal current or wattage, STOP! This isn't right. Leave the Variac set well below this point, or at a point where the circuits are drawing enough power to give you some readings. Now, get out the dc voltmeter, and start checking some bias voltages, etc. Something is definitely wrong in here, and it will show up in the form of incorrect bias, no dc voltage (due to a short) where there should be some, and so on. By holding the applied power down to a level where you are not drawing enough current to blow out the new transistors, you can protect the amplifier until you find the additional trouble. Incidentally, one of these Variacs costs just a little bit *less* than ONE 50-watt power transistor. See? If you can save even one of these expensive devices, you'll have paid for the Variac in a hurry.

This isn't the only use for this piece of equipment. In many of the smaller phono amplifiers, the ac power comes from a tap or secondary winding on the phono motor. If the motor is burned out, and you want to check the amplifier to see if it's in working order, simply disconnect the motor winding, connect it to the Variac, and turn it up to supply the ac voltage you need. This will be somewhere around 25 volts; check the schematic to be sure.

You probably have a bench dc power supply for checking transistor radios, auto-radios, etc. Many of these have a fixed output, 12 volts, dc, etc. You can make this a fully variable dc power supply by plugging its line cord into the Variac. Start at zero, and bring the line voltage up until you get the voltage you need. You can even go slightly above the normal output if need be; most Variacs will raise the voltage by about 15 to 20 percent above the applied ac line.

In low-voltage dc power supply circuits using full-wave bridge rectifiers or voltage doublers, the versatile Variac can even be used as a substitute for a burned-out power transformer. Any type of transformer that does not have a center-tapped secondary can be checked with this. Just unhook the bad transformer and clip the output leads of the Variac to the rectifiers. The ac voltage can be set up to whatever it was before, and the amplifier tested to see if there is anything else wrong; in other words, the eternal question: Did this power transformer break down internally and burn up, or was it burned up by an external overload? The Variac can give you the answer.

CRISSCROSS-CHECKING STEREO AMPLIFIERS

Stereo amplifiers are a great invention, especially for the service technician. If one channel is dead or weak, which is a fairly common complaint, you can use the working channel as a reference and make A-B checks between the same test points; dc voltages, signal levels, etc., can be compared. This comes in handy in cases where you can't find a schematic diagram. You'll often find both channels on the same PC board; one channel will be a "mirror-image" of the other, making it a little easier to trace the circuits. Tie both inputs together, set the balance control in the middle, and then cross-check between channels to get the information you need.

Comparison of power output in both channels can be made by turning the gain or loudness controls to the same point and measuring the ac voltage developed across the speaker voice coil. Power can be computed by the Ohm's law formula $P = E^2/R$, using the speaker impedance as R. Comparison of distortion in either channel can be made by checking the output waveform of each channel, with identical input.

CHAPTER 5

Very Small Amplifiers

Up to now, we've discussed all of the basic circuits that you will find in this type of amplifier. We've also covered the necessary test equipment and test methods. From here on, we'll take up typical examples of commercial amplifiers, both mono and stereo, and show their peculiarities, their circuits, and how to check each of them. In a great many cases, you'll find a special use of a piece of test equipment that will make testing this particular unit easier. You'll also find some special devices in a few makes and models that will be, basically, just the same as some of those explained before. We'll show you how these work and how to test them in the quickest possible way. While they may be the same, they won't look like it!

We'll start out with the smallest transistor amplifiers. These are used in small record players and have an output of about 500 milliwatts. From these, we'll go on through the medium sized and bigger amplifiers, and then on to the real window-breakers found in musical instrument amplifiers and even in some home music systems. Some of these have (quite literally) a thousand times as much power output as the tiny phono amplifiers. One of the big units to be covered is an RCA amplifier with a power output of *500 watts*, which is exactly 1000 times the 500-milliwatt (0.5 watt) output of a phono amplifier!

SMALL MONAURAL PHONOGRAPH AMPLIFIER

Only a few years ago, small audio amplifiers were very "perceptible." They had power transformers, output transformers, and metal chassis, and weighed 5 or 6 pounds. Now, they have almost disappeared! The power and output transformers have disappeared, tubes are replaced by transistors, and the whole "works" will be assembled on a PC board, often only about 2 inches square. If it weren't for the volume and tone controls, which are still normal size, the amplifier would be hard to find.

The PC board is usually mounted on a thin sheet-metal bracket, which also holds the tone and volume controls. These controls go through holes in the cabinet, and sometimes hold the amplifier to the cabinet. Speakers vary in size, but are usually bigger than the amplifier itself.

Fig. 5-1 shows the circuit of a typical miniature transistor amplifier as used in a small mono record player. This is probably the most

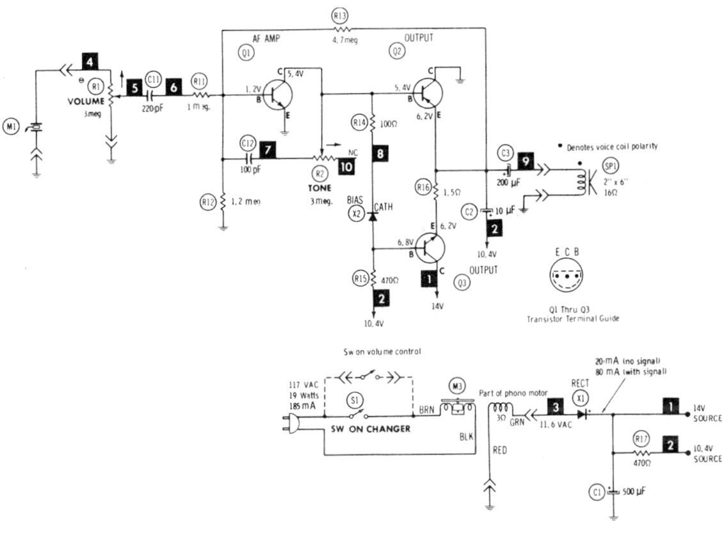

Fig. 5-1. Typical small phono amplifier.

common use for these tiny amplifiers. The PC board for this amplifier is only 1.6 by 3.1 inches, and holds all of the parts except the speaker and controls. Only three transistors are used. The output stage is a complementary-symmetry pair, direct coupled to a single driver transistor. A bias diode is used, in the base circuit of the lower transistor.

(NOTE: although you may find the *diagrams* inverted, from now on the "top" transistor in an output pair will always be the one to which the dc voltage is applied, and the "bottom" transistor the one that has its collector or emitter grounded. The diagram is inverted in Fig. 5-1; note the dc voltage going to the lower transistor

(Q3). "Top" and "bottom" refer to the initial dc voltage level (see Chapter 2).

No phase inverter is needed with this type of output circuit. The bases are tied together, and the driver feeds in a full sine-wave signal. Each of the output transistors uses only its "forward-bias" half-cycle. All of these mini amplifiers use the output-transformerless output circuit, to save space, money, and parts. Power output of a typical circuit like this one will be, at a generous estimate, only about 500 milliwatts.

The dc power supply is a simple half-wave rectifier. The ac supply, at 12 volts, comes from a secondary winding on the phono motor. A 500-μF electrolytic capacitor does the filtering. A dropping resistor reduces the +14-volt supply to 10.4 volts for the driver transistor. An extra 10-μF electrolytic capacitor helps to prevent feedback in this circuit.

Parts Location and Service Data

Fig. 5-2 shows a bottom view of the PC board, with the location of the transistors and various test-points called out. This is taken from a Sams PHOTOFACT Folder, 1030-8. As you can see, there is no information printed on the underside of *this* board. Later on, you'll find that many of the amplifiers built in the U.S. have an "overprint" on the wiring side of the PC board, with the part numbers, transistor bases, and other helpful information. The cheaper imports seldom have this very useful feature.

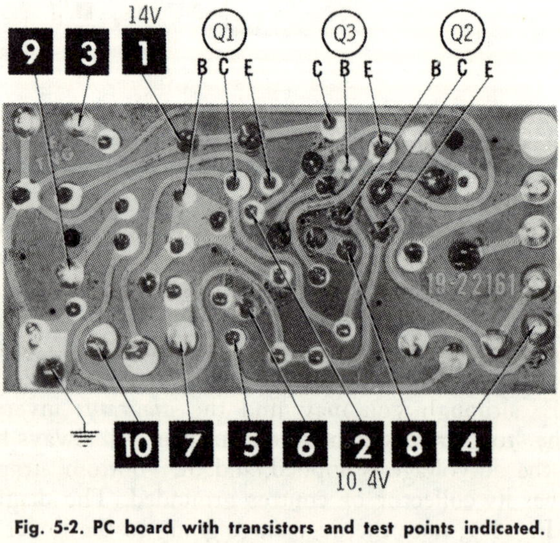

Fig. 5-2. PC board with transistors and test points indicated.

So, you can see the importance of having ample service data when trying to make repairs to this kind of equipment. However, in the very simple mini's, it won't be too hard. After all, this amplifier has a total of 16 parts, not counting the speaker. That's right—16: six resistors, two ceramic capacitors, three electrolytic capacitors, two controls, and three transistors. Fortunately for us, the circuitry will be pretty well standardized; they're all going to look and work a great deal like this one. Variations will be only in the types of transistor used, npn or pnp, and the difference between the stacked (identical output transistors) type, and the complementary-symmetry (similar but opposite polarity transistors) type.

Dc supply-voltage points are called out on the photo layout. However, you can locate them on the board itself by finding two key parts, the rectifier diode and the input filter capacitor. The junction of the rectifier and the capacitor will always be the highest dc-voltage point. If this is a positive voltage supply, the cathode of the rectifier and the positive terminal of the capacitor will be connected. If it's a negative voltage supply, the rectifier anode and the negative terminal of the capacitor will be tied together. In this example, the 14-volt source is "Point 1" on the "*CircuiTrace*" (the black blocks with figures). These are shown on both the board photo and the schematic. Point 2 is the low voltage, 10.4 volts.

Testing

Of course, the dc voltage supply should be the first thing checked. After this, check the dc voltage at the midpoint of the output pair. Here, this would be on the emitter of Q3. In this circuit, you should read about +6.2 volts. Remember, this test will work with all types of OTL transistor circuits. This should be very close to half of the applied dc voltage on the top transistor. ("Top": the one to which the dc goes first.)

If the midpoint voltage, zero volts, is missing, the top transistor, Q3, is probably open. If Q3 is open, no dc voltage can be applied to the output capacitor, so there will be no midpoint voltage. This is the most common failure in these amplifiers. In a few cases, trouble in the driver transistor or bias diode has caused the top transistor to be biased to cut-off. Here, the driver voltages can be checked, since they come from a separate supply. A very high negative voltage on the base of Q3 indicates that it is cut off. Alternate check: turn off the power, and check the emitter-collector junction of Q3 with the ohmmeter. Look for the diode effect. If you get an open-circuit reading both ways, the transistor is open. If you get a normal diode-effect reading, the transistor could be good but biased off.

A high dc voltage at the midpoint could indicate a collector-to-emitter short in Q3. The dc voltage could go on through, and charge up the output capacitor. No signal, of course. A bias defect that caused the base of Q3 to go highly forward biased (more positive) would cause this transistor to go into saturation (conducting maximum current) and show the same symptom. Quick check: read the base and emitter voltages of Q3.

An open bias diode will reduce the bias voltage on Q2, but will not make it cut off, since the driver transistor collector has a separate supply. It will cut off the signal from Q3, however, and result in a very high distortion. The scope will catch this type of trouble instantly.

Small upsets in the midpoint voltage, such as reading 8 or 9 volts, or 3 or 4 volts instead of the normal 6.2 volts, could be due to leakage in the driver transistor Q1. This must be taken out of the circuit in order to get a valid test. The junction of Q3 which is shunted across it will make leakage tests impossible. If the base voltage of Q2 is quite a bit off, suspect the driver transistor first.

Signal Tracing

This type of amplifier is so small that it might seem hardly worth-while to make signal-tracing tests. However, the scope can help you pin down any distortion and spot troubles that would be otherwise pretty difficult to locate. Any problem that causes a very severe distortion or loss of volume will be very apt to cause dc-voltage upsets, and can be caught in that way. The scope will always be helpful, though, so don't hesitate to use it in any of these tests.

You can make an overall test of the whole system by using a sine-wave test record on the phonograph. Block the player up level with empty parts boxes or blocks of wood, and play one of these records, checking the output with the scope. This will check the amplifier and the phono cartridge, stylus, etc., all at once. There are many different test records, and they're very useful things to have around. Later on, in the better grade amplifiers, we'll be using them to check frequency response of the systems. Many of them have "frequency runs" with bands of frequencies from a very low to very high, all recorded at the same level. With this, a quick check of the overall response of the whole system can be made with little trouble. Frequency runs are seldom necessary on this type of amplifier; they're not noted for their "Hi-ness of the Fi!"

MINI-AMP WITH DARLINGTON DRIVER

Fig. 5-3 shows what looks very much like an ordinary amplifier with a standard circuit. The output is; it's a complementary-symmetry pair, just as in previous units. However, though the driver transistor is *drawn* as a single-unit, it is a special type. It is actually

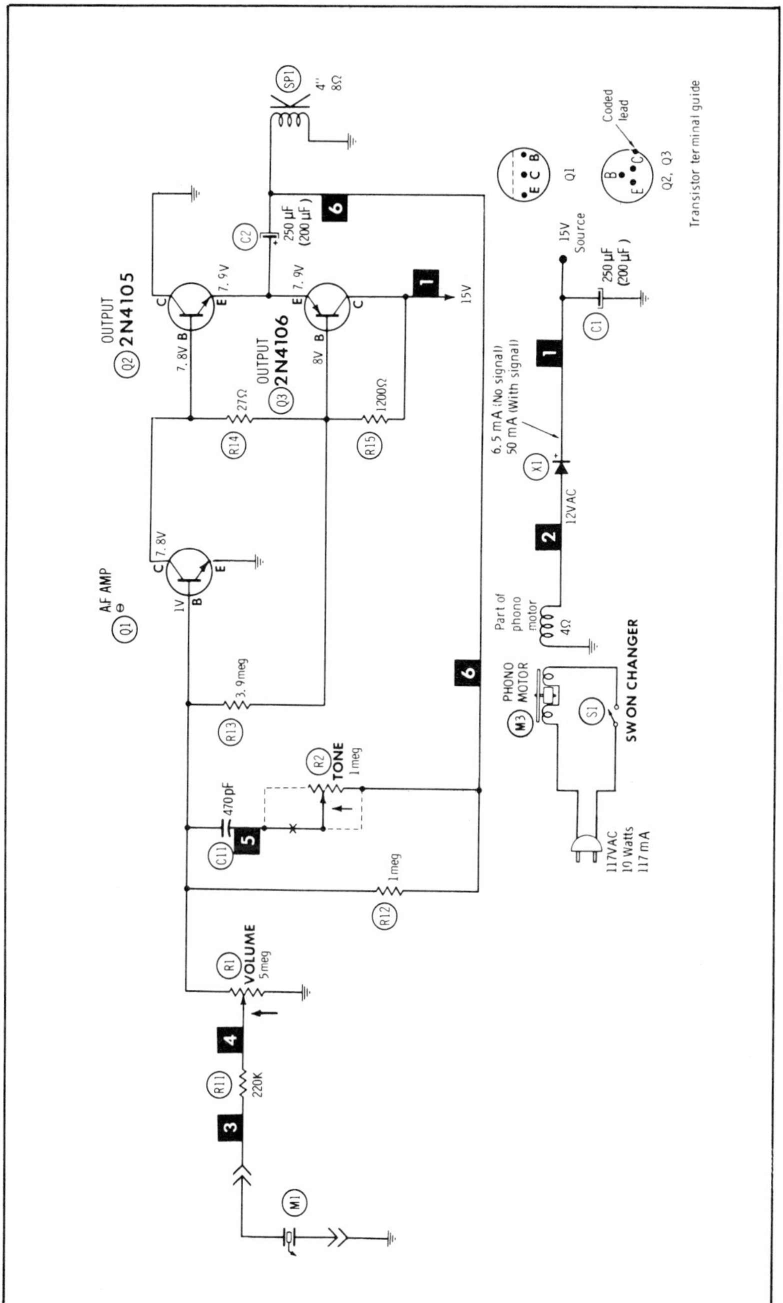

Fig. 5-3. Circuit of mini-amp with Darlington driver.

two transistors built in a single case, and connected in what is called a "Darlington-pair" circuit. This has several advantages over the stock single-transistor driver, so let's take a look at this circuit and its characteristics.

You'll probably be finding these in quite a few amplifiers later on. One major manufacturer is building "power Darlingtons" in a single case; this looks like the standard TO-3 case from above, but from underneath, it doesn't; it has 8 connections! This can almost make up a whole amplifier by itself.

The Darlington Pair Circuit

Fig. 5-4 shows the connections of a Darlington pair. The dashed line represents the case of the compound unit. As you can see, there

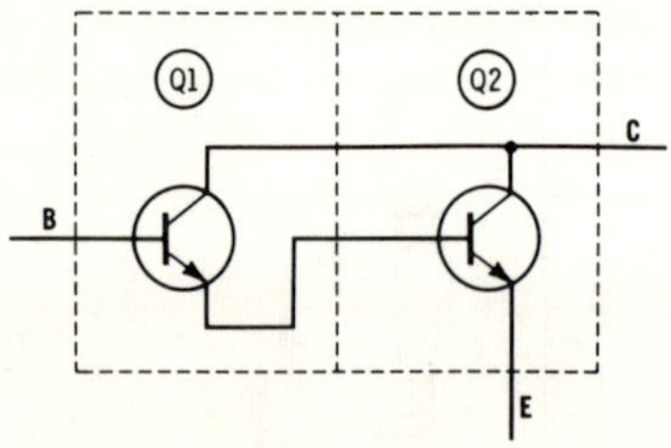

Fig. 5-4. Single-unit Darlington pair, in one case.

are only three terminals brought out: base, emitter, and collector. So, the external connections are the same as a regular transistor. This is where the confusion can come in!

Actually, this is a two-stage direct-coupled amplifier. The two collectors are tied together internally, and a connection brought out. The emitter of transistor Q1 goes directly to the base of transistor Q2—no external connection. The emitter of transistor Q2 comes out as the emitter of the combination.

Transistor Q1 is actually working as an emitter-follower, driving transistor Q2, which is a common-emitter. This type of combination is capable of an extremely high forward-current transfer, meaning very high gain. An emitter-follower has a high current gain, and a common-emitter has high voltage gain. So, the two, together, can give a much higher amplification than they would if used in the conventional circuitry.

The Darlington also has a much higher input impedance than the common-emitter unless the latter is bootstrapped (which takes more parts; so—). Low input impedance has been one of the biggest headaches in transistor circuit design. Most of our regular "input devices," such as microphones, phono cartridges, guitar pickups, and so on, are inherently high-impedance devices. By using a Darlington as a preamplifier, we can get the benefits of high impedance input for better matching, and a higher gain, which is always desirable since it reduces the number of stages (and parts).

I went into detail on this, so that you won't fall into the trap that several of us have, and replace this unit with a single stock transistor! Then, you sit there and wonder what happened to the *gain* of the record player. You'll find a note in the parts list telling you that this is a special Darlington-type transistor unit. In cases of absolute emergency, two separate but identical transistors could be used. Tie the collectors together, make the floating connection from Emitter 1 to Base 2, and then install them in the regular holes. Any standard general-purpose audio transistor ought to be pretty close, same polarity and type, with a beta of say 50 or so.

I made up one of these, just to make sure that this idea *would* work. In the amplifier of Fig. 5-3, the Darlington driver was taken out and replaced with a pair of HEP-53s. Dc voltages and signal levels were checked and logged first. The HEP-53 is a silicon transistor with a 30-volt collector rating and a beta of 85. The leads were twisted together to make the Darlington connection, and it was tacked into the circuit.

Scope tests of the waveform showed no distortion. Pk-pk signal voltages were exactly the same. A listening test confirmed these results. However, I did get a variation in the dc voltages. The Darlington base voltage went up 0.1 volt, from +1.0 volt to +1.1 volt. The collector voltage rose from +7.8 volts to about +9.8 volts.

Since this circuit is direct-coupled, and the Darlington collector voltage is the base voltage of the output transistors, it might be better to use transistors with a slightly lower beta, if you're making a "permanent" replacement. This can be verified by reading the collector current of the output pair, and checking it against the value shown on the schematic. If the bias change is too great, raising this current more than a small amount, try a different pair for the driver. In this case, the 2.2-volt increase in forward bias on the output did not seem to have any bad effects. I believe it would be best to hold this bias change down to not more than 1.0 volt, if possible; the collector current will be the final deciding factor, of course. Be sure to check for waveform distortion and loss of volume, since *this* is what the customer will be listening for!

As a final experiment, I substituted a single HEP-53 for the Darlington, and got just what I expected (which surprised me somewhat!). The gain was very low, and the distortion was very severe. The collector voltage and output bias went quite a bit higher, up to more than 11.0 volts. This probably accounted for the distortion, due to changing the crossover point of the output pair. However, the gain was completely unacceptable, and that alone was enough to reject it.

Testing of Darlington Pairs

In direct-coupled circuits, which are probably the most common type, you can't make in-circuit tests on the Darlingtons, or on the output transistors either, for that matter. Too many shunt paths. So, they'll have to be taken out of the circuit for tests.

On the standard types of transistor testers, such as Sencore TF-151 and similar units, you'll get some "peculiar" results, although you can make a valid test of the units if you know how to interpret the results. A good Darlington will show no leakage. However, it won't "Calibrate" on the beta test. It will probably read somewhere about half-scale, on the Low-power ×1, or Low-power ×10 setting. When the "Test" button is pushed, though, the needle will go off-scale to the left, indicating a very high beta.

This kind of reading is probably due to the stacked or direct-coupled internal connection of these units. The test voltage may not be high enough to drive them into conduction as it does with single transistors. Due to time limitations, I didn't have time to verify this, but it sounds likely. At any rate, this was verified with my home-made Darlington, and a couple of others. So, if you get this type of reading, it is probably all right.

The low CAL reading is probably due to the fact that we have actually got two transistors "in series" or cascade. With silicon transistors, the "on-bias" of 0.6 volt each adds up to 1.2 volts. The typical transistor tester does not apply a very high voltage to the transistor; so, the cascaded pair of a Darlington won't conduct. If you try this with *germanium* transistors, which need only 0.2 volt each to make them conduct, they will CAL, and read a very high beta; much greater than the sum of the betas of the two transistors tested separately. Takes only 0.4 volt for conduction. However, so far as I know, this circuit is not being built with anything except silicon types, so the previous statements will apply.

The use of dc voltage and scope tests in-circuit and a leakage test out of circuit will be the final verification. Of course, an open Darlington will check just like any other open transistor. There's a quick check you can make for opens; hook the unit to the transistor tester, set on the Leakage setting. In this case, an NPN setting should show no leakage; turning the polarity switch of the tester to PNP should show full-scale deflection. If you get no reading at all in either position, the unit is open. (Or one of the clips has fallen off the transistor leads!) This quick test for opens is useful for all transistors, by the way.

If you make up a substitute pair and you find the dc voltages, signal waveform, etc., pretty close to the original, the chances are that it will work very well. If you can not find a substitute pair that will do this, then you will have to order an exact duplicate Darlington from the factory distributor.

If you get an amplifier with very low gain and high distortion and the output transistors are all right, check the Darlington! If you take this out and find a HEP-53 or similar replacement transistor, somebody has goofed! If this transistor checks perfectly good, it's probably a mistaken replacement. By the time you finish this book, replacements for these transistors will probably be available from the replacement-transistor manufacturers. (See the next item.)

Fig. 5-5 shows a stereo amplifier of the same general type. The major difference between this and the preceding circuits is only that there are two of them. A complementary-symmetry output circuit is used, with transistors of the same class. Here again, a Darlington driver is used, a type 16P3367.

HEAT SINKS

When replacing the output transistors in this type of amplifier, there are two things you must always do. One: As mentioned before, use exact-duplication replacements, matched-pairs, for either the stacked or complementary-symmetry types. The replacements should have the same cases as the originals. The next paragraph tells why.

Two: Be very sure that the heat sinks are correctly replaced. The output transistors used in these amplifiers are normally pretty small, and are driven very hard to get as much power as possible from them. So, for maximum transistor life, heat sinks must be used to keep them as cool as possible.

You'll find two different types of heat sink used. If the very small epoxy-cased TO-92 types are used, the most common heat sink is a flat plate of aluminum, about one inch square. A spring clip holds the flat side of the transistor case against the heat sink. Fig. 5-6 shows a top and side view of these. The transistor case fits up into a little notch in the bottom of the plate, and a metal spring clip holds it tightly. Since the cases of these transistors are insulated, the heat sinks are not electrically "hot." Silicone grease is used on most of them, to improve the heat-transfer ratio from case to heat sink. Transistors *must* fit very tightly to the heat sinks.

If the round metal-cased Type TO-5s are used, you'll find two different types of heat sink. In many of these transistors, the collector is electrically connected to the case, once again to raise the heat-dissipation ability. So, the case will normally be hot. Either the case or the heat sink must be insulated.

One type of heat sink is a large flat metal plate, usually aluminum, mounted at right angles to the PC board. This plate can be seen in

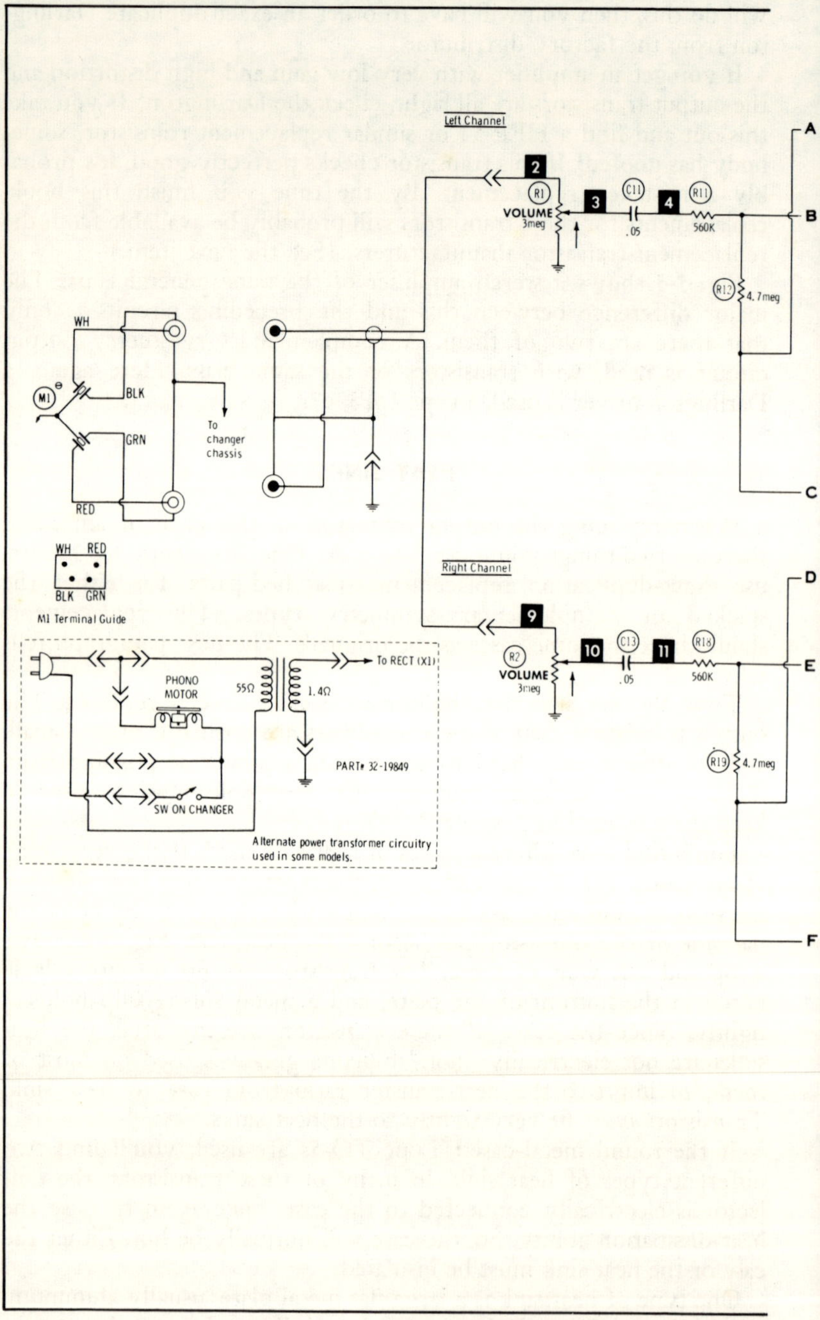

Fig. 5-5. Stereo amplifier with Darlington

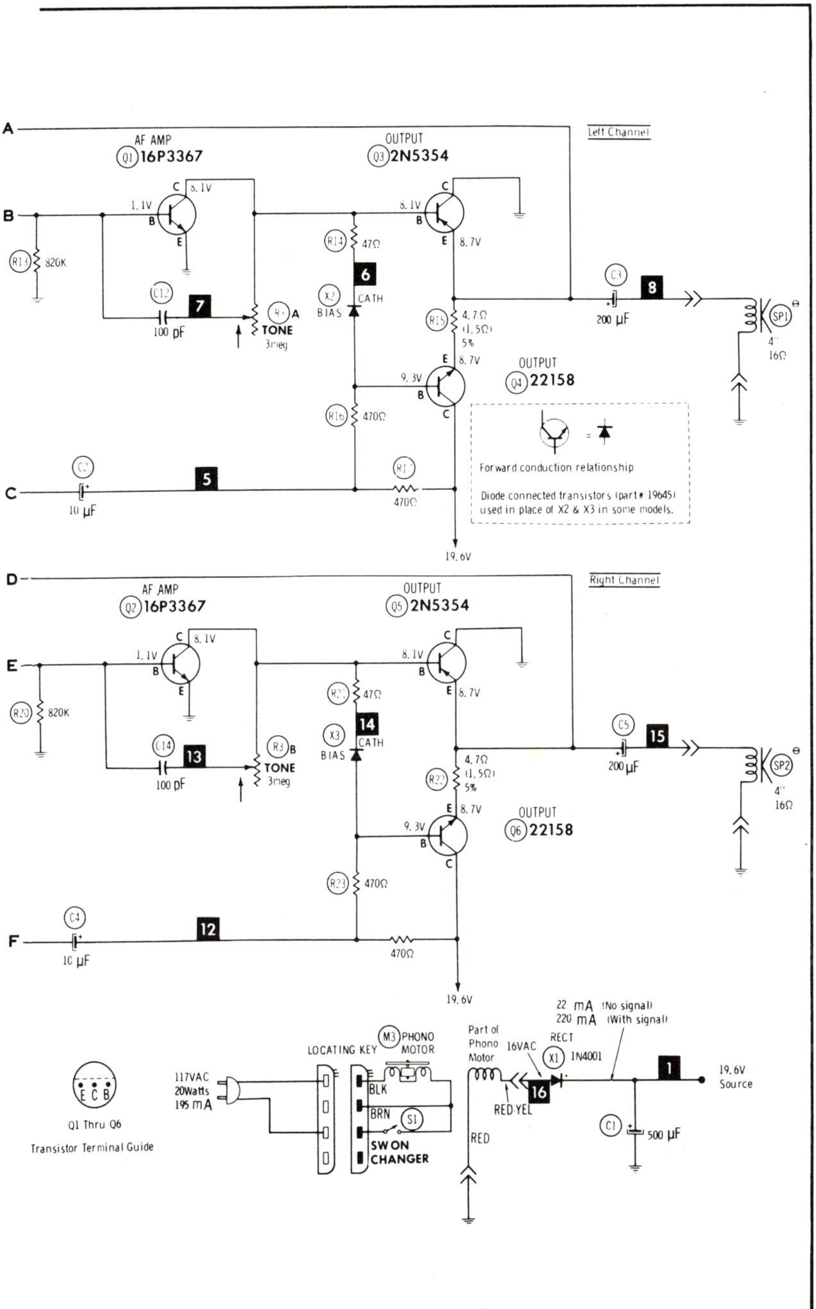

driver and complementary-symmetry output.

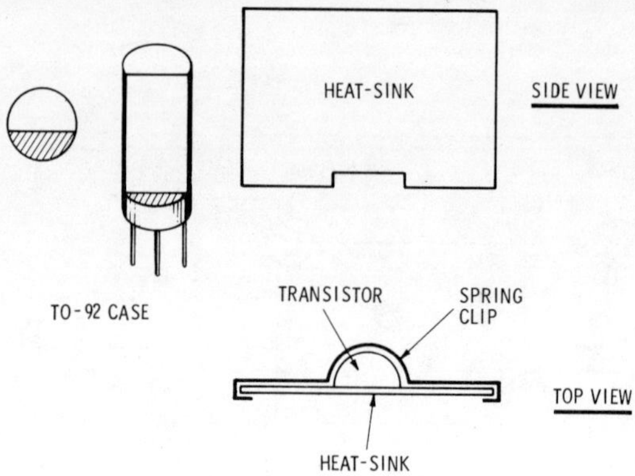

Fig. 5-6. Clip-on heat sinks for TO-92 case transistors.

the photo of Fig. 5-7 (the amplifier of Fig. 5-3). Checking the schematic, you can see that the collector of Q2 is grounded. So, this transistor can be clamped directly to the heat sink. The collector of Q3 gets the full dc supply voltage, +15 volts, so this one must be insulated. You'll find a mica insulator between case and heat sink, also between case and clamp.

When you install one of these transistors with sink, leave the base and emitter leads open until the transistor is mounted. Make an ohmmeter check between transistor case and heat sink, to make SURE that the collector is not grounded. If the original mica in-

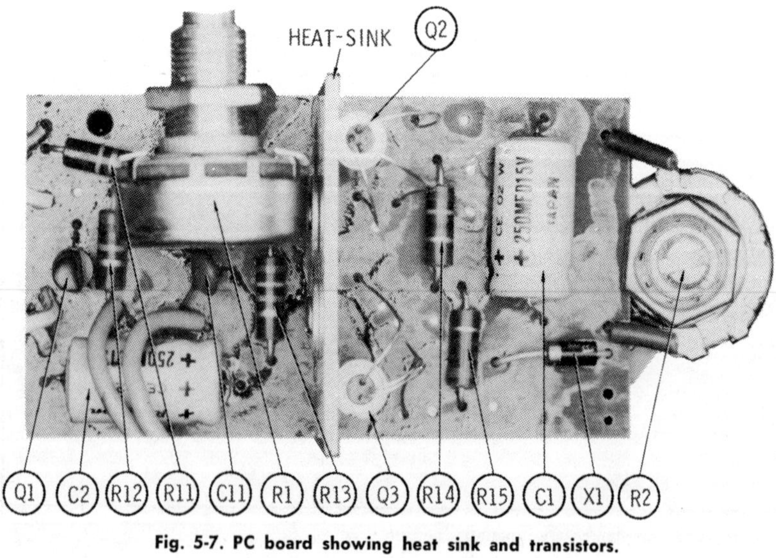

Fig. 5-7. PC board showing heat sink and transistors.

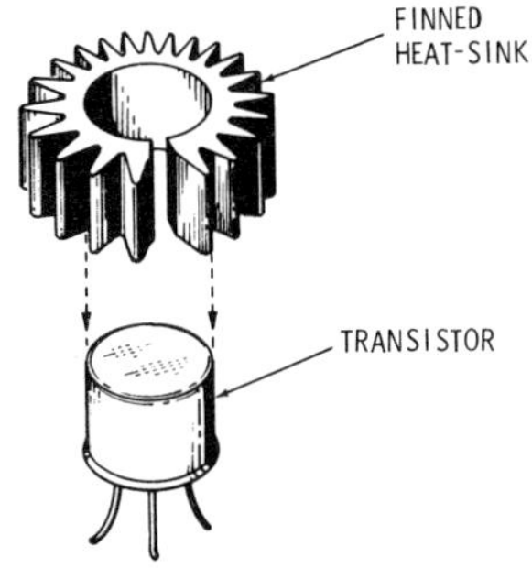

Fig. 5-8. "Push-on" heat sink.

sulators have been cracked or broken, use new ones. You'll find these packed with the new transistors in most instances. If not you can improvise good insulators from scrap mica. (One good source of mica for this is the insulation from a discarded electric iron, toaster, etc. This can be obtained at electric appliance repair shops. You can get sheet mica in several different sizes.) Use silicone grease between the transistor case and clamp, etc. This helps the heat transfer as well as the insulation.

The second type of heat sink used with the TO-5s is sometimes called a "push-on." These are made of flat sheet metal, usually painted black, or of cast aluminum. Fig. 5-8 shows their construction. These heat sinks must fit tightly over the transistor case for best heat transfer. If they are hard to get on, you can use the silicone grease to help get them into place. This type is ordinarily not insulated from the transistor, so the heat sinks will be electrically hot, and must be kept clear of grounded objects and other transistors and parts in the vicinity. If the original transistors had heat sinks, as shown by the photos in the service data, but they're not there now, get new ones. Several different makes can be found at radio-tv supply houses.

In case you wonder why the emphasis on the heat sink, run one of these amplifiers for about 30 minutes, then feel the output transistors! This will put an end, once and for all, to any ideas you may have had that "transistors run cool!" They do not run as hot as tubes, of course, but they do run pretty hot. Later on, when we get into the bigger units, you'll find output transistors that can raise a nasty blister on an incautious fingertip.

SERVICING

Servicing of these amplifiers follows the same general method outlined in previous chapters. Use the 3-step method: dc power supply first, then output transistors, then preamps and drivers. With an input signal of any kind, even a regular phonograph record, you can follow the signal through the circuit until you find out where it stops. On a musical record, (instrumental, rather than vocal) you can even get a fairly good idea of clipping! The waveforms won't be stationary, of course, but you'll see a decided "flat-topping" of the pattern. This will produce a brighter line, with the peaks going up from it. Normal signal will show peaks on both top and bottom. All stages of these amplifiers up to the output are Class-A, and the signal across the speaker, of course, should be normal again.

Dc voltages should be *close* to those given on the schematic. Contrary to some of the original statements, you'll find that dc voltages on transistors are NOT quite as critical as they were supposed to be. In one sense, that is—bias voltages are critical, with respect to the emitter, as we've said. But, if *all* dc voltages on a given transistor are low or high by the same amount, you'll find a surprising lack of difference! Most of the significant dc-voltage differences will be large enough to be obvious; you'll find the full supply voltage on a point where there should be at least a 50-percent drop, or zero voltage at a point which should read about 15 volts, and so on, so that any dc-voltage upset that is great enough to cause trouble won't be too hard to find.

Contrary to another "well-known fact," most of the troubles you'll find in these amplifiers will be defective transistors. They are not immortal, as we were first led to believe! So, make them the first things in a circuit to be checked. After you have pinned the trouble down to a given circuit, be sure that the transistor itself is good. In-circuit transistor checkers are very reliable, if their readings are properly interpreted. If they show that the transistor does have a certain beta, then the chances are that it is good—at least, good enough to pass a signal. If it shows the transistor to be "shorted," don't believe this—*not yet*. Take the transistor out of the circuit and recheck for shorts. A short or leakage reading means more now because there's nothing else across the transistor.

Shorted parts in the circuit can simulate a shorted transistor. For example, a shorted bypass capacitor from base or emitter to common return will give this impression. This will upset the bias, or kill the signal, and give a very good imitation of a shorted transistor. Even in this case, taking the transistor out and testing it will be the fastest way. Statistically, you'll find more bad transistors than you will other parts. So, check the transistor first. If it shows good, then leave it out, and make resistance checks from each element to ground or common, to find out if the parts left in the circuit really *are* good. For example, if you see a 56-k bias resistor from base to ground but your ohmmeter shows only 2000 ohms, look for a shorted capacitor.

CHAPTER 6

Medium-Powered Amplifiers

Having discussed low-powered amplifiers, we now get to the next class, which is the medium-power types. This includes amplifiers from about 5 watts output up to perhaps 25 or 50 watts. Once again, you'll see the same old circuits: output-transformerless output circuits of all types, practically all of them in Class B, and a few circuits using input transformers. Some circuits use output transformers, with tapped secondaries. This last category is found mainly in standard public-address system amplifiers, where various combinations of speakers must be used. This flexibility in speaker loading is an essential for a good pa system. The only difference between these, and the amplifiers you have seen so far, is in the greater power output, and the use of more preamplifier stages to get the drive needed.

POWER OUTPUT RATINGS

Power output ratings, always a source of controversy, have by now been at least partially standardized. In the past, a lot of manufacturers have rated amplifiers at "peak power," meaning the absolute maximum that the amplifiers could deliver. This caused confusion and interminable arguments. The purists complained that this kind of rating was misleading, which is quite true. No amplifier can deliver its full peak power without getting into an area of high distortion, mostly third harmonic distortion. You can demonstrate this for yourself. Turn the volume control on any amplifier up to about 75 percent of maximum, and listen. The output should be loud, but clear and "clean." Turning the volume control on up to maximum, you'll hear only a small increase in absolute output, but you will note a very decided increase in distortion.

Now, at least a partial agreement has been reached on power-amplifier ratings. Many manufacturers now rate amplifiers in terms of "music power." This is a rating worked out in conferences by the Institute of High Fidelity and the EIA. You'll sometimes find this called "IHF power." The official definition of this term is "the power measured immediately after the sudden application of a signal, and during a time interval so short that power-supply values within the amplifier have not changed from their no-signal values." If you want to translate that into more usable terms, just think of it as the power output at which the amplifier has less than 5 percent distortion (with a continuous signal such as music).

This does bring down the ratings, at that. An amplifier capable of delivering 40 watts of peak power now has only a 20-watt rating! For our purposes, this tempest in a hi-fi teapot is neither here nor there. What we want to do is get the amplifiers back into condition so that they *can* deliver that 20 or 40 watts of power.

WHERE YOU WILL FIND THEM

In this chapter, we are dealing only with the audio amplifiers. From here on, you may find the audio amplifiers used alone, as pa systems, musical-instrument amplifiers, etc., or you may find them used in combinations, with am/fm radios, record players, or tape players, etc. Even then, you can still separate the audio amplifiers from the other circuitry. They will be separate units in the majority of cases, with their own power supplies. Also, they provide many of the cases of trouble, due to the high power consumption. So, they can be taken up separately.

Fig. 6-1 shows a typical audio system of this type. This one is used in several different models; one with a record player only, another with am/fm radio, another with radio plus record player, and so on and on. The IHF power rating of this unit is 7.0 watts per channel, for a total audio power output of 14.0 watts. Remember, in stereo systems the "total power" rating means the sum of the power output of both channels.

Note the output circuit. This uses a stacked Class-B circuit, with identical pnp transistors, driven by an input transformer. Here, for the first time, we see the use of a dual-voltage power supply. The output transistors are supplied by a +18-volt and −18-volt dc source. With a separate dc voltage supply for the "bottom" transistor, in this case the one with the negative voltage applied to its collector, the "charging action" of the speaker coupling capacitor is not needed.

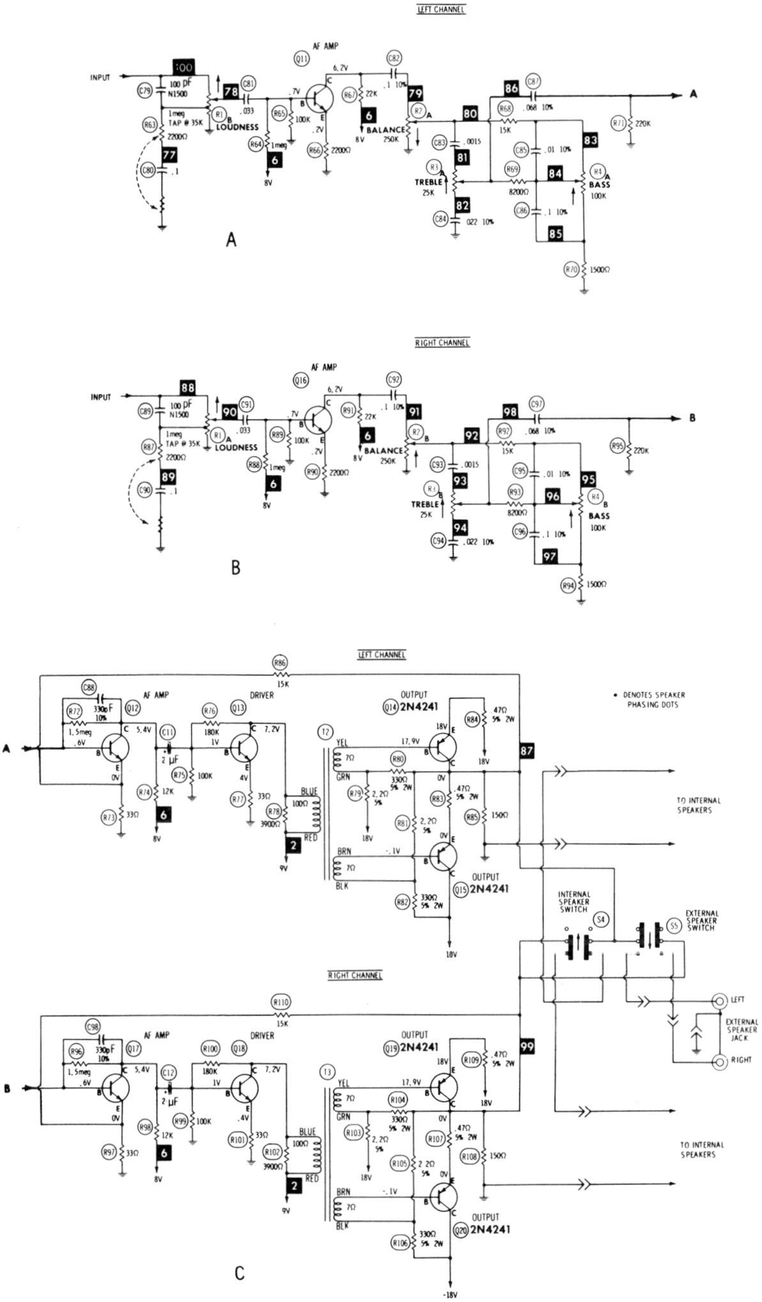

Fig. 6-1. Stereo audio amplifier circuit.

So, you will note that the speakers are connected directly between the midpoint of the circuit and ground.

So, now we have a change in the "typical characteristics" of the circuit (for test purposes only; the *circuit actions* are exactly the same as the previous ones). We'll get some different dc voltage readings. The dc resistance of the speaker voice coils will be pretty low, from 5.5 ohms to 8 ohms, etc. So, you will not read "half the supply voltage" at the midpoint of this circuit. The key clue here should be the presence of a positive voltage on the emitter of the top transistor, as before, BUT there will also be an equal but negative voltage on the collector of the bottom transistor. In the previous circuits, this point was ground—remember?

The dc voltage at the midpoint should be zero, especially with no signal, while the transistors are resting. In some versions of this circuit, there may be a small voltage here to offset the tendency to distortion during crossover. However, in the typical circuit of this type, there should be practically no *dc* present here. There will be a good-sized ac (signal) voltage present, of course, which you can check with the scope. Once again, the scope is the fastest and most definite way of making sure that the circuit is working properly.

THE DOUBLE DC POWER SUPPLY

As you can see from the schematic of Fig. 6-2, the output stage is supplied with dc voltages, equal but opposite in polarity. This is easy to do. A center-tapped secondary winding is used, with two simple half-wave rectifiers, reversed to give the needed polarity. A

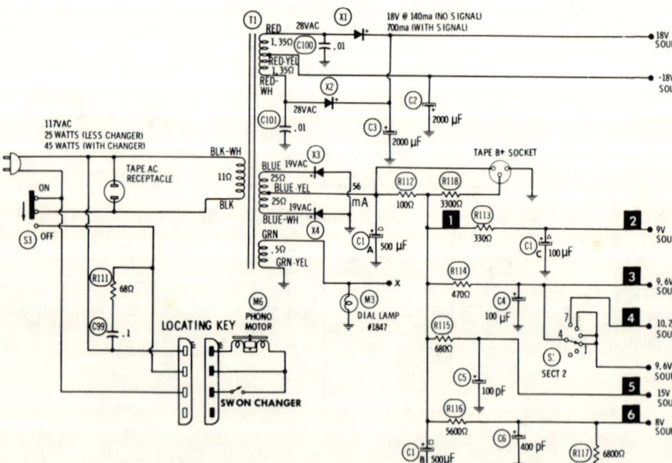

Fig. 6-2. Power supply with positive and negative outputs.

pair of 2000 μF electrolytic capacitors provide the filtering for this part of the supply. This is the high-current winding.

Preamplifier and driver stages are supplied from a standard full-wave rectifier circuit, with a lower current rating, below. Note that the diodes here have their anodes grounded, with the positive dc voltage taken from the center tap of the secondary winding. This is a little unusual, but it works just the same. Other voltages are taken from dropping resistors. Actually, only the 8.0- and 9.0- volt sources are used to feed the audio amplifier; the rest are used in the radio circuitry. If this were the straight audio type, these sources would be omitted. They are separately filtered. Large electrolytics and the dropping resistors give better isolation for the small-signal circuits, to avoid feedback and/or oscillation.

Notice the current readings shown on the dc output terminals of these power supplies. These are a standard feature of Sams PHOTOFACT service data, but are too often omitted from the manufacturer's service data. They can be very useful indeed, when you are checking for proper performance of the dc power supply, the output stage, or for any kind of fault which could be in the power supply itself or elsewhere in the circuitry.

For example, take a case where the 8-volt source is low. We have two possibilities: one, the 8-volt source is "weak"; it is not able to provide the normal amount of load current to the load. Check for open input filter capacitor, or similar trouble. Two, there is an overload, short, or high leakage somewhere in the *load* circuits. Which one is it? To get the answer in the shortest possible time, break the circuit at the filter input; in this one, at the center tap of the transformer secondary. Connect a dc milliammeter in series, and turn the power on again. The current you read will tell you instantly which of the two troubles you have.

Normal current here is shown as 56 mA. If the meter indicates anything close to this, say between 50 and 60 mA, it's normal. (However, the current won't be normal; you've got a fault in there somewhere, remember?) If there is far too much current, this means that the power supply itself is in good condition. If it is capable of supplying far *more* than the rated current, it's all right. The trouble must be somewhere in the load circuits. You've eliminated half of the problem.

If the current is far too low, this means that the cause of the trouble is in the power supply itself. Even with this reduced current, it is not able to hold the *voltage* up to normal. So, the load circuits can be considered to be all right for the moment anyhow, until the power supply is checked out. Something like an open input-filter capacitor or an open rectifier diode in a full-wave or bridge circuit can cause this type of symptom.

Note that a single test has already given you a valuable clue as to the location of the fault, and has saved a lot of checking of perfectly good parts. You've traced the fault to either the load or supply section by one simple current reading.

If a defective silicon diode is found in any of these circuits, be very sure to check the *current* rating of any diode used for replacement. Up to this point, we're still working in low-voltage type supplies. Anything up to, say, 50 volts is considered as low voltage. The stock silicon diodes of the type used for tv power supplies will always have ample voltage ratings. However, some of the older types may have only a 0.5 ampere (500 mA) current rating. Check the maximum current drain shown on the +18- and −18-volt supplies. This is 700 mA, at full output. So, any replacement diodes used in this circuit must have a minimum of 1.0-ampere (1000 mA) current rating. If they're not high enough, they'll run very hot and blow out within a very short time. If a 1.0-ampere diode looks big to you, just wait a little while. In the following chapters, you'll see diodes that must have ratings of *20 amperes* and more.

THE BIG TRANSISTORS

From now on, you'll be running into a different type of power transistor, used in the output stages. These amplifiers have much greater power output, so the power transistors must be able to handle much higher currents. They must also be capable of dissipating the rated power within the transistor itself. In such a circuit, just as in a tube power amplifier, part of the power is dissipated in the load, as useful power output, and the rest is dissipated in the transistor or tube. Fig. 6-3 shows this action. This is the reason for the popularity of Class-B circuits. In these circuits, each transistor is dissipating only 0.25 to 0.312 of the full power. In Class A, the transistor dissipates a much higher percentage of the power. So, Class-B circuits are much more efficient.

There is a possibility of an "area of confusion" here between the power-dissipation rating, per device, and the power *handling* ability of the transistor. All of the audio power that appears across the load (the speakers, R_L in Fig. 6-3) must go *through* the transistor(s). So they must be able to handle this much load and carry this much current. The power-dissipation rating of the transistor itself means the amount of power that is used up in heating of the transistor junction, case, etc. This is actually *wasted*—used up inside the transistor, and *not* delivered to the load. This rating will be found in all transistor data charts, replacement guides, etc., and must be checked before selecting any replacement power transistors. More details on this in Chapter 10.

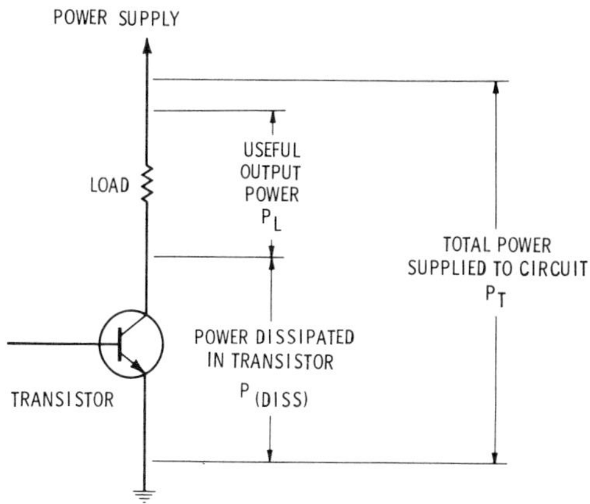

Fig. 6-3. Power distribution in a transistor stage.

THE TO-3 CASE POWER TRANSISTOR

Fig. 6-4 presents four types of transistor cases not previously shown in this text. The TO-3 will be used, in one form or another, for all of the high-power types from now on. It is roughly diamond shaped with a dome on top for the "works," and has two holes on the ends for mounting with screws or bolts. In every one of these examples, the case is electrically connected to the collector. This gives these transistors a much better thermal transfer ratio to get rid of the heat developed across the junction (the P_{diss} we just mentioned). The heat flows from the collector to the case, from the case to the heat sink, and is then radiated off into the air. The purpose of this design is to keep the junction well below the critical temperature where it could go into thermal runaway and destroy itself.

This is the reason for the large heat sinks you'll find, and the liberal use of silicone grease between the transistor and the heat sink. The grease provides a much better thermal coupling, especially in the common case where the transistor must be insulated from the heat sink with a thin mica insulator.

Most of the older power amplifiers used germanium transistors. These are more prone to thermal runaway, and require larger heat sinks. Their critical junction temperature is around 85 to 110 degrees celsius. (Temperature in degrees C = 0.556 times temperature in degrees Fahrenheit, minus 17.8.)

The silicon power transistors have practically taken over, especially in the high-power field. They have a much higher critical temperature, from 150° to 200° C, and special types can be made that will go up to 300° C.; they can run considerably hotter and

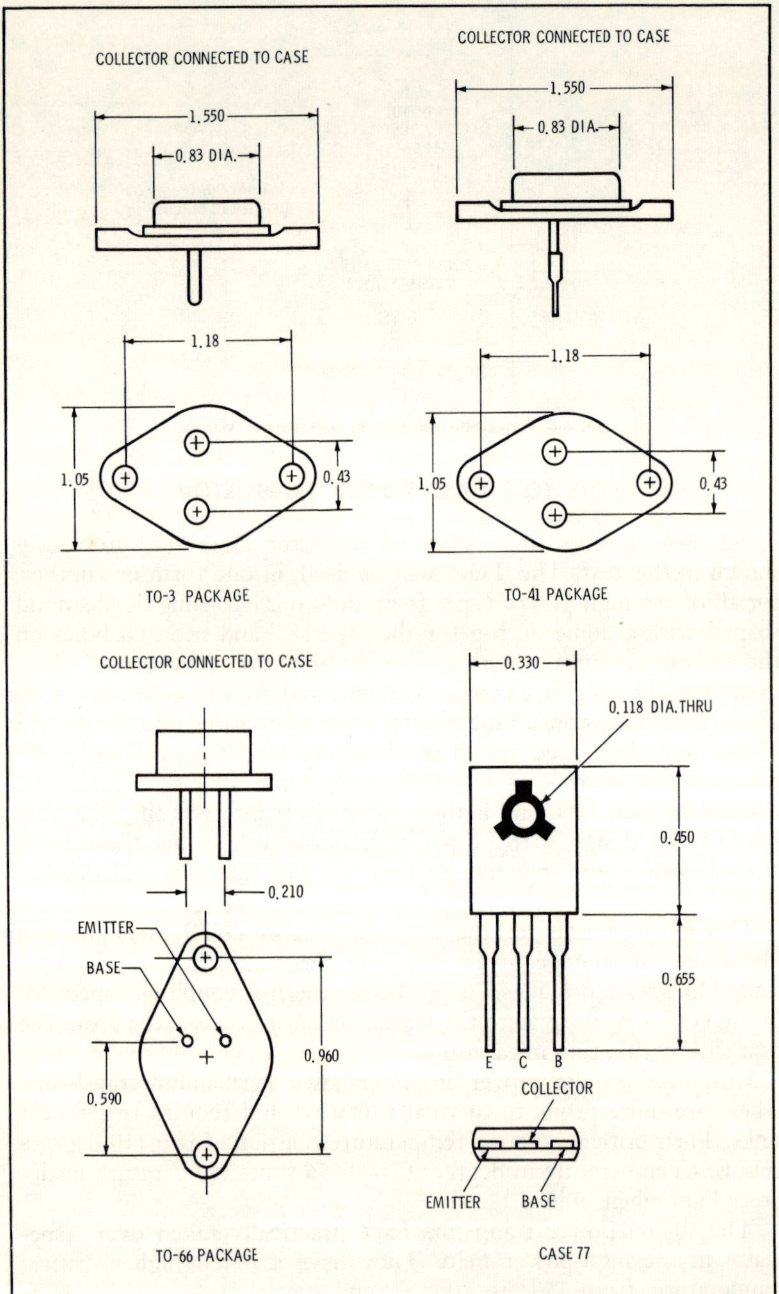

Fig. 6-4. Transistor cases and connections.

survive, and they do. *Don't* touch the power transistors of a big amplifier after it has been running for a while. Some of them can get hot enough to cause painful burns.

The TO-41 case is physically the same as the TO-3. Only the base and emitter pins are different. In emergencies, these can be used interchangeably with TO-3s, provided the electrical characteristics are suitable. In smaller amplifiers you'll find another case that looks like these, but is smaller. This is the TO-66 case. Transistors built in this style are generally rated lower than the bigger TO-3 and TO-41 types. A new case is flat, with leads coming out of the end, and is called "Case 77." The physical size of this case will allow you to replace TO-66 type transistors with it, provided that electrical characteristics match. More on this later.

CASES AND MOUNTINGS FOR TO-3 TYPES

The mounting of these big transistors is of a special type, and must be done very carefully to avoid several possibilities for trouble. Fig. 6-5 shows an exploded view of this type of mounting. The case is held to the heat sink by two bolts or screws through the holes in the ends. There are two variations in this mounting. In one, the bolts are threaded into the metal of the heat sink and must be insulated from the transistor case (collector) by insulating shoulder and sleeve washers.

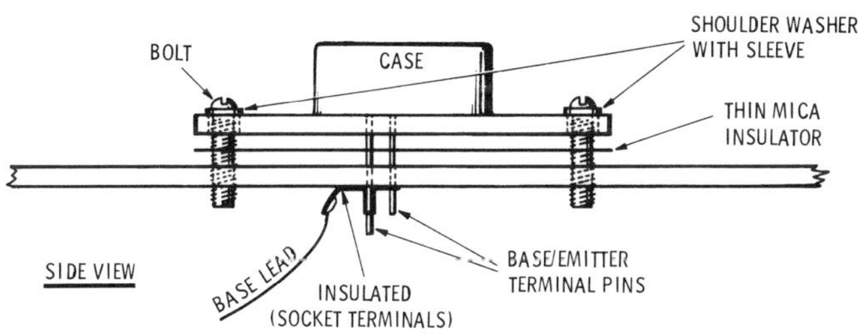

(A) Side view.

(B) Top view.

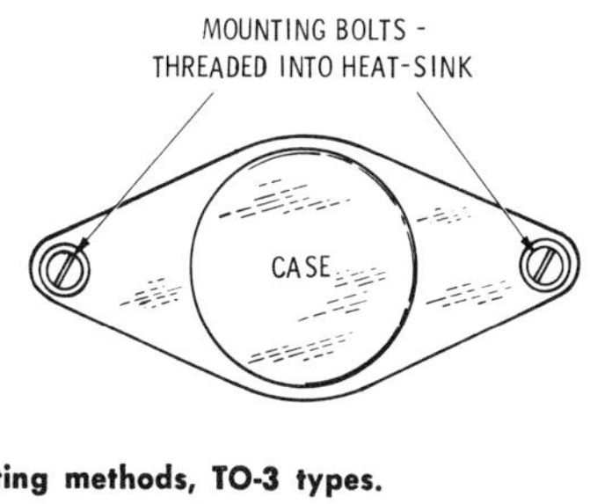

Fig. 6-5. Mounting methods, TO-3 types.

In the second version, insulated spring clips are mounted to the underside of the heat sink. The bolt holes are large enough to provide clearance and keep the bolts from shorting to the heat sink. Self-tapping bolts go through the mounting holes of the transistor, with no insulation, and thread into the spring clips (Tinnerman clips) under the heat sink. This has one very definite advantage for us, as we'll see soon.

In both of these mountings, the base and emitter are connected to rigid pins; holes in the heat sink are large enough to provide clearance. Spaghetti or other insulation is sometimes slipped over the pins to avoid shorts. In the second type of mounting, terminals (like tube socket contacts) are provided on the insulating wafer. This type allows the transistor to be "plugged in" and bolted down; no unsoldering is needed. In sets with cheaper construction, you'll often find the base and emitter wiring simply soldered to the pins, a method that saves money.

The plug-in type of mounting has very definite advantages for us. From now on, we're going to have to take collector current readings once in a while to make sure that allowable maximum-current ratings are not being exceeded, and so on. If the plug-in type of mounting is used, we can do this with ease, as Fig. 6-6 shows. Note that Bolt No. 1 goes into a clip underneath the heat sink and that there is also a terminal lug for the wiring to the transistor collector. The electrical connection from the wiring to the collector is made by

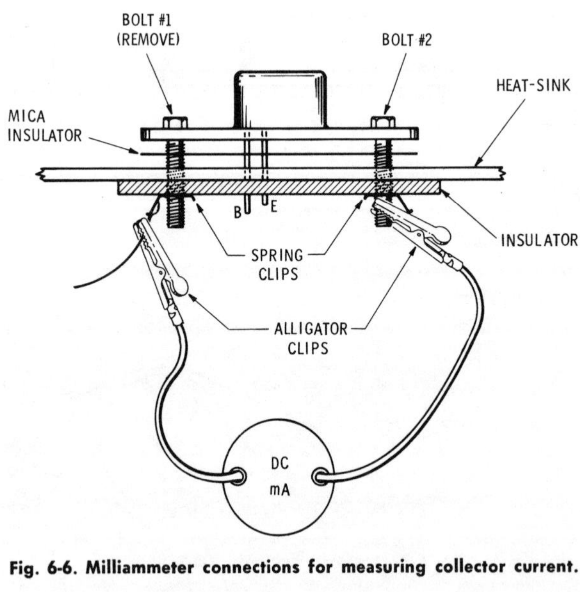

Fig. 6-6. Milliammeter connections for measuring collector current.

this bolt, since it is shorted to the case. (If this one is left too loose, it can be the source of some very interesting intermittents, too. Remember that!)

To take current readings, all we have to do is take out Bolt No. 1. Now, if we clip our dc milliammeter between the two terminals as shown, source and collector (which is still connected to Bolt No. 2), the milliammeter is in series with the collector. Now we can read the collector current of this one transistor, all by itself. This type of measurement is also very handy for checking resting currents, or for matching currents when making transistor replacements, and so on.

HEAT SINKS

We haven't bothered to draw the actual size or shape of the heat sinks in the preceding illustrations, because there is such a wide variation. In medium-powered amplifiers, they will be perhaps 2 inches square. In bigger amplifiers, they will be much larger and may have fins to increase the area of surface exposed to the air. This makes them radiate the heat much faster.

In a few cases, you'll find the transistors bolted directly to the heat sink—no insulation. The whole heat sink will be insulated from common or ground. This is often easier than insulating the transistors, and possibly improves the heat-transfer ratio because of the direct contact between case and heat sink.

In such instances, the heat sinks will have the full collector voltage on them. If this is fairly high, and you'll find collector voltages of 100 volts and up in the bigger ones, it can be a shock hazard. In all instances, it can be a *short* hazard. Make sure that nothing else, especially any grounded metal object, can either fall on, be bent into contact with, or touch the heat sink or transistor cases. This will blow the fuse, at the very least, and the resulting transient spike can (and has been known to) blow the transistors.

UNUSUAL OUTPUT CIRCUITS

You're going to find quite a few variations of the standard stacked or complementary-symmetry circuit. Here are a few of them. Don't let them throw you—they all do the same thing in the same way, but there are a few minor differences in connections.

First is a circuit such as that of Fig. 6-7. This is a medium power amplifier, probably designed or use in radio-tv stations or recording studios, etc. The difference here is that the speaker is not driven directly from the midpoint of the output stage to ground. The primary of a matching transformer is connected at that point, instead.

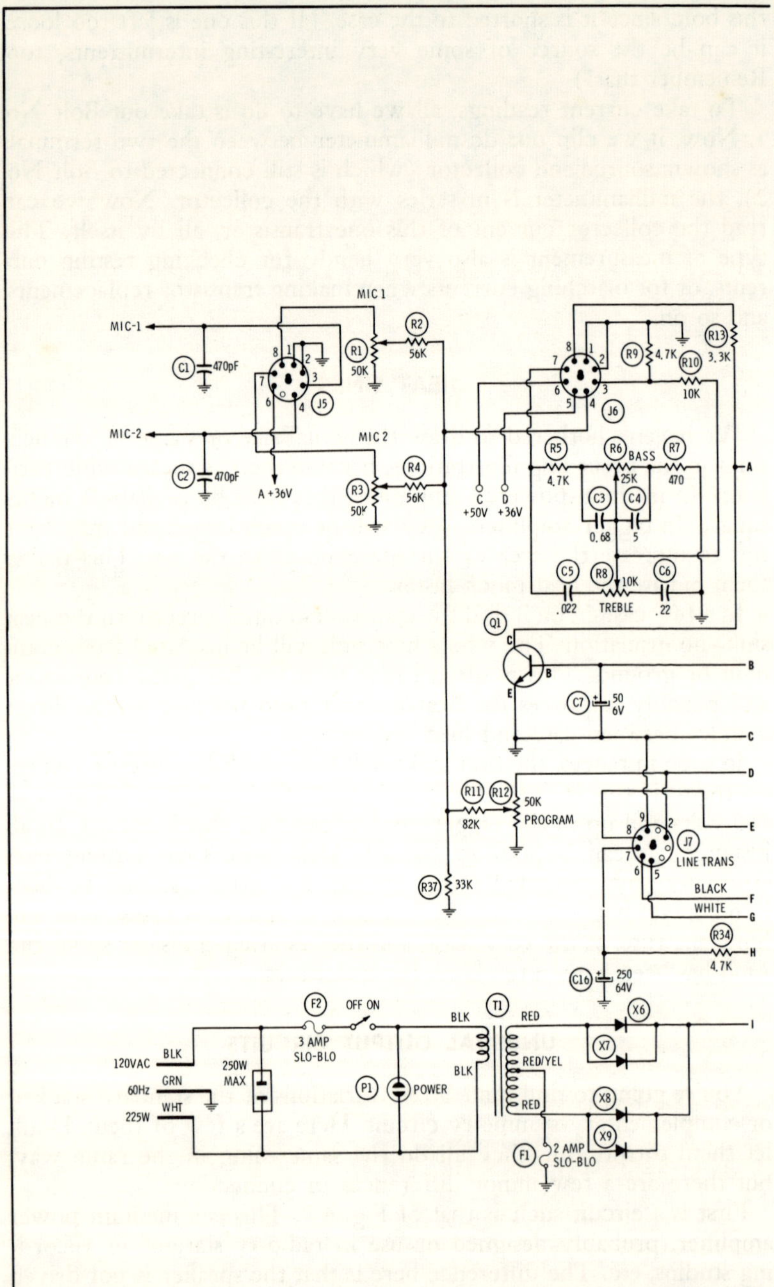

Fig. 6-7. Transformer-coupled amplifier

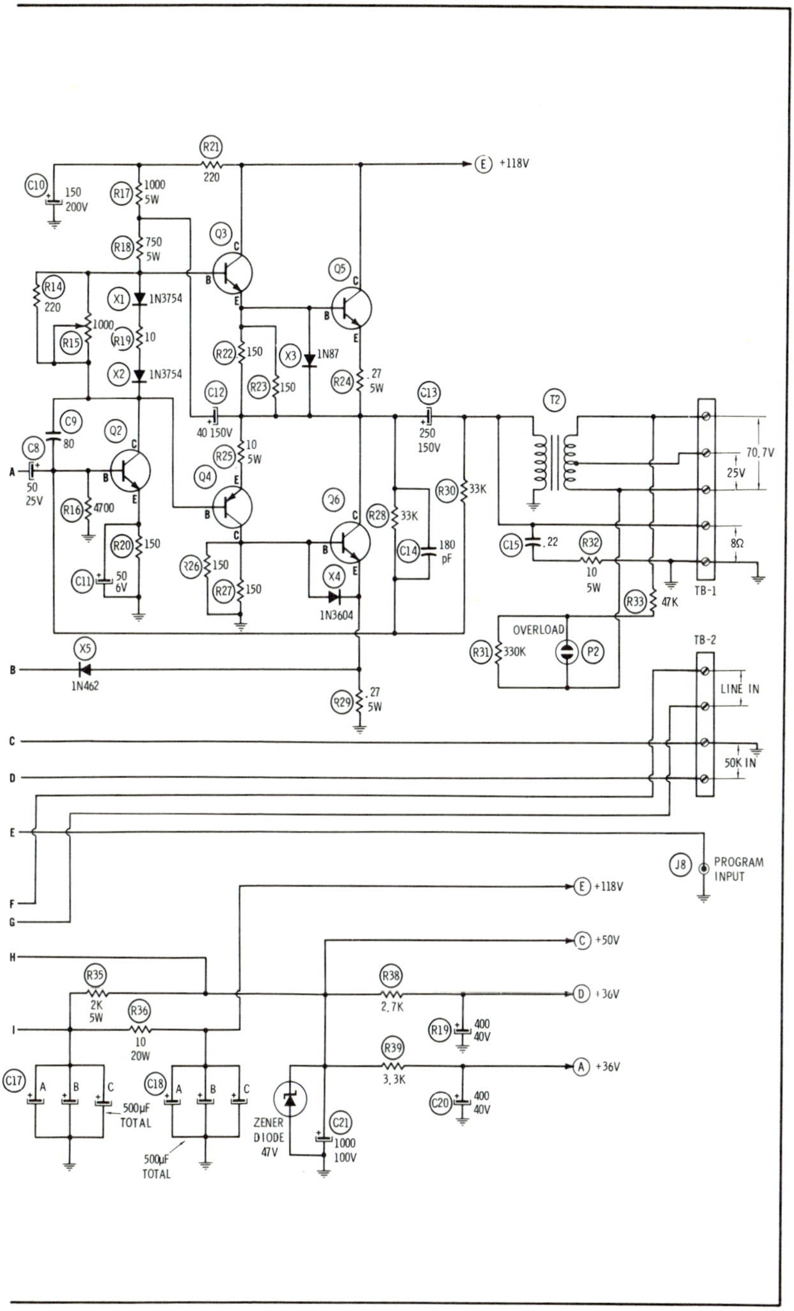

for radio/tv station use.

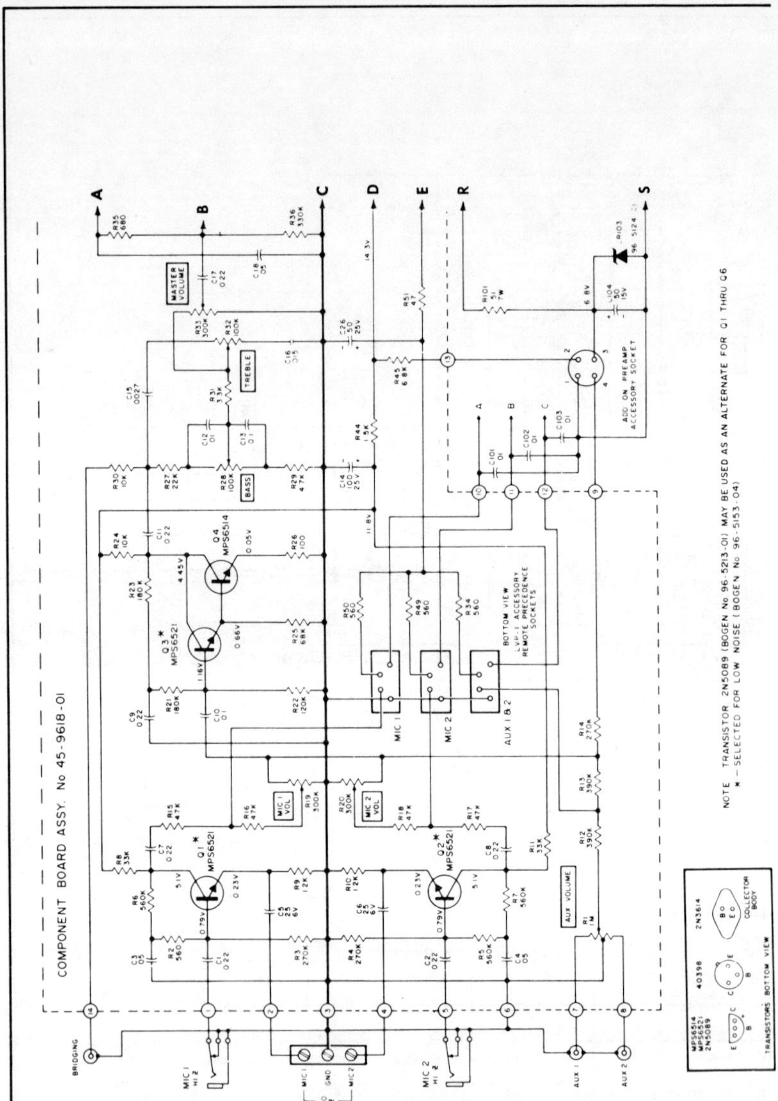

Fig. 6-8. Bogen CHS35

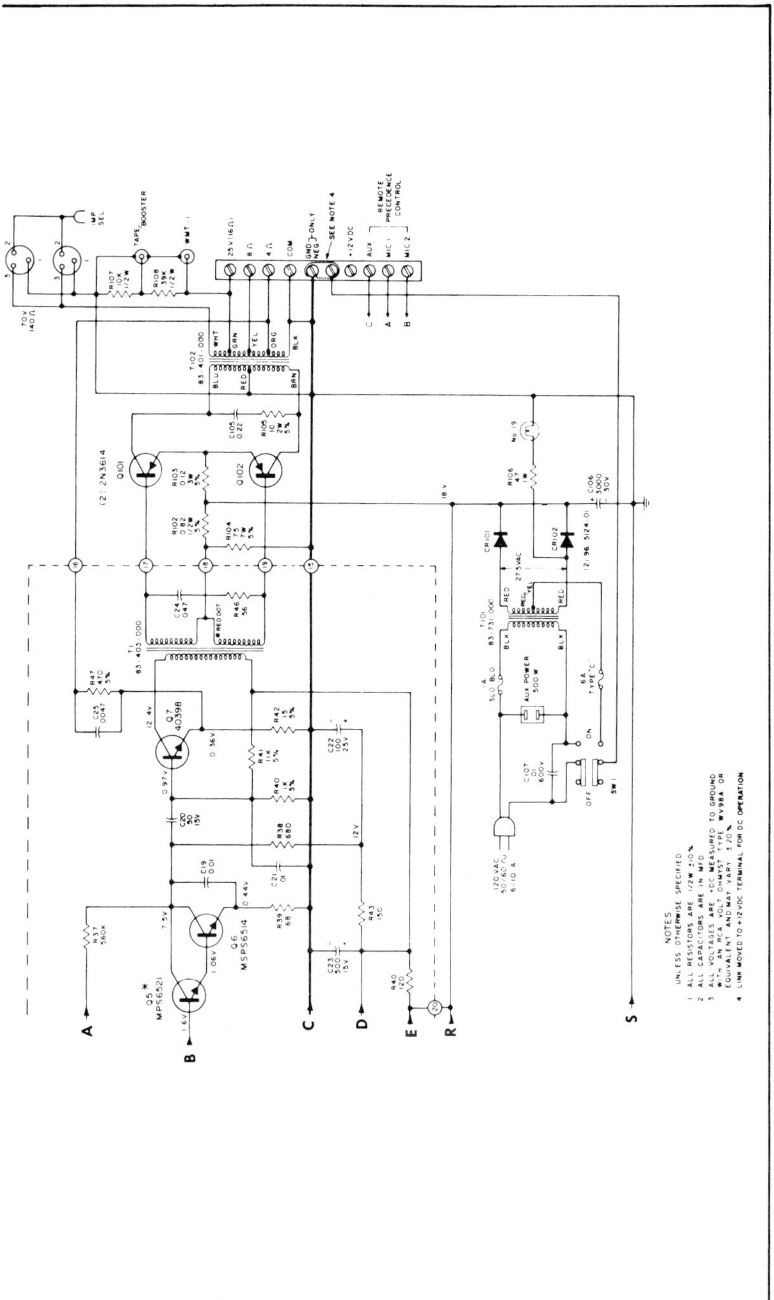

amplifier for PA system.

This is a transformer with secondary windings to drive either a 70.7-volt or 25.0-volt line.

These are called "constant-voltage" lines. Several different speakers can be driven from this line, if the proper matching transformers are used between each one and the line. Different amounts of power can be taken off as needed. The total power output of the amplifier is thus divided among all of the speakers, in amounts determined by the impedance ratio of the matching transformers. Total power output of the amplifier cannot be more than its normal rating, of course.

Here, a high dc voltage is used; note the collector voltage of +118 volts. High rated transistors must be used as replacements, with ample collector voltage-breakdown rating. As an oddity, note that the top transistors, Q3 and Q5, actually form a Darlington pair. The output transistors, of course, are Q5/Q6, a stacked pair, driven by a complementary-symmetry pair, Q3/Q4, to get the necessary phase inversion, etc. Bias and clamp diodes are used to hold the bias constant with temperature changes.

PUBLIC ADDRESS-TYPE AMPLIFIERS

Fig. 6-8 shows the circuit of a Bogen CHS35 transistorized public-address system amplifier, rated at 35-watts output. Note that input and output transformers are used. At first glance, this looks exactly like the old push-pull circuit used so often with tubes. It does *work* in the same way, but there are differences. If you'll follow the power supply lines to the output pair, you'll see that their emitters are tied together and are connected through R103 to the +18 volt source in the dc power supply.

One collector is tied to each end of the output transformer primary. However, instead of going to "B+," as in the tube circuits, the primary center tap goes to ground. There is actually a small negative voltage on the collectors; the path to ground is through a protective pilot lamp, a 47-ohm resistor, half of the power-transformer secondary, then to ground through the on-off switch. So, this point is effectively at signal-ground. The No. 19 pilot lamp acts as a protective device. Like all incandescent lamps, its filament has a highly nonlinear resistance as the current varies. If a high surge of current is drawn by the output stage for any reason, the lamp filament will absorb it by lighting up. Of course, if too much current is drawn, the filament will burn out; so, it also acts as a fuse.

Despite differences in polarity, etc., this circuit works exactly the same as any push-pull tube circuit. Signal paths are the same. The output transformer has a multiple-tapped secondary, which is an essential in any amplifier intended for pa work. This allows the use of a wide range of speaker combinations, as long as the impedance is matched to that shown on the terminal board.

Fig. 6-9 shows the circuit of a Bogen CHS50, a 50-watt, solid-state, pa system amplifier. At first glance, the output circuit looks exactly the same as the preceding one, but there are differences. In this circuit, note that the *collectors* are tied together on the output pair, and connected directly to the +35-volt dc source. The output transformer primary is center-tapped, as before, but now its ends are connected to the emitters of the transistors. The center tap is directly grounded. Since the signal current is the "emitter-collector" current, it doesn't make much difference *where* we put the load.

Taps on this primary winding, looking a lot like the old "Williamson" amplifier circuit, go back to the bases, through the input-transformer secondary windings. This is a feedback network, to correct for possible distortion and allow a greater undistorted output.

Overload protection in this amplifier is provided by a standard circuit breaker in the primary of the power transformer. The No. 19 pilot lamp, with a 220-ohm resistor in series, seen above the transformer, is just that—a pilot light.

EVER-PRESENT HELPS IN TIME OF TROUBLE

Now, we're getting into an area where things are more complex. Not because they are different; the same circuits are used, but now there are more of them. These are all on PC boards. With the wiring on one side and the parts on the other, circuit-tracing can be very difficult unless we have some way of knowing just where each part is. Of course, you *must* be able to find parts with the greatest of ease, to make signal tracing and voltage measurement tests, etc.

We are getting a lot of help from *some* manufacturers, in addition to the Sams PHOTOFACT Folders. Fig. 6-10 shows the factory schematic of a Zenith 10ZT31 audio amplifier, 16 watts output stereo. Notice the stage-gain figures shown above each circuit. This can be a great help in finding out just which stage isn't working. As we said before, there can be stages that actually show a voltage loss, and here is a good example; note the −8 dB figure shown from the base of Q402 to its collector. This is a good-sized loss, but it is normal, since this stage *is* a current-amplifier. By contrast, the next stage, the driver, Q403, will show a gain of +48 dB. The output stage shows a slight loss, −1 dB. Remember, these are *voltage* figures. These readings are the easiest to make, since they do not require opening of circuits; so that makes it very easy to check.

Sams PHOTOFACT provides a phantom view of the underside of PC boards for parts. Fig. 6-11 shows this for the same amplifier. Note

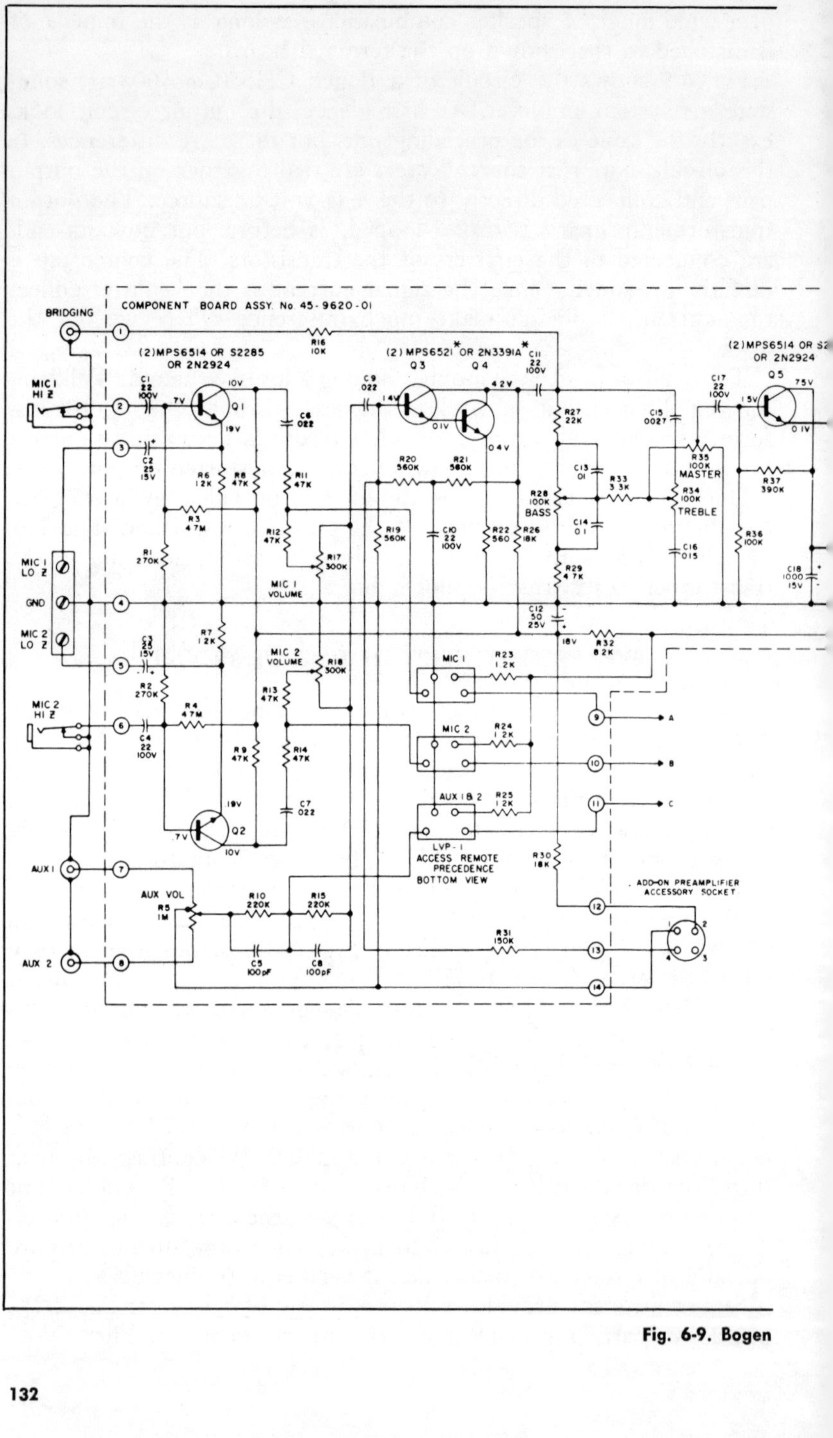

COMPONENT BOARD ASSY. No. 45-9620-01
BRIDGING
MIC 1 HI Z
MIC 1 LO Z
GND
MIC 2 LO Z
MIC 2 HI Z
AUX 1
AUX 2
(2)MPS6514 OR S2285 OR 2N2924
(2) MPS6521* OR 2N3391A*
Q1
Q2
Q3
Q4
Q5
(2)MPS6514 OR S2
OR 2N2924
R16 10K
C1 22 100V
C2 25 15V
C3 25 15V
C4 22 100V
C6 .022
C7 .022
C9 .022
C11 22 100V
R6 1.2K
R8 47K
R11 47K
R3 4.7M
R1 270K
R12 47K
R17 300K
MIC 1 VOLUME
R7 1.2K
MIC 2 VOLUME
R18 300K
R13 47K
R2 270K
R4 4.7M
R9 47K
R14 47K
R20 560K
R21 560K
R19 560K
C10 22 100V
R22 560
R26 18K
R27 22K
C13 .01
R28 100K BASS
C14 .01
R29 4.7K
C12 50 25V
R33 3.3K
C15 .0027
R35 100K MASTER
R34 100K TREBLE
C16 .015
C17 22 100V
R37 390K
R36 100K
C18 1000 15V
R32 8.2K
MIC 1
R23 1.2K
MIC 2
R24 1.2K
AUX 1 & 2
R25 1.2K
LVP-1 ACCESS REMOTE PRECEDENCE BOTTOM VIEW
R30 18K
ADD-ON PREAMPLIFIER ACCESSORY SOCKET
AUX VOL R5 1M
R10 220K
R15 220K
C5 100pF
C8 100pF
R31 150K
A
B
C

Fig. 6-9. Bogen

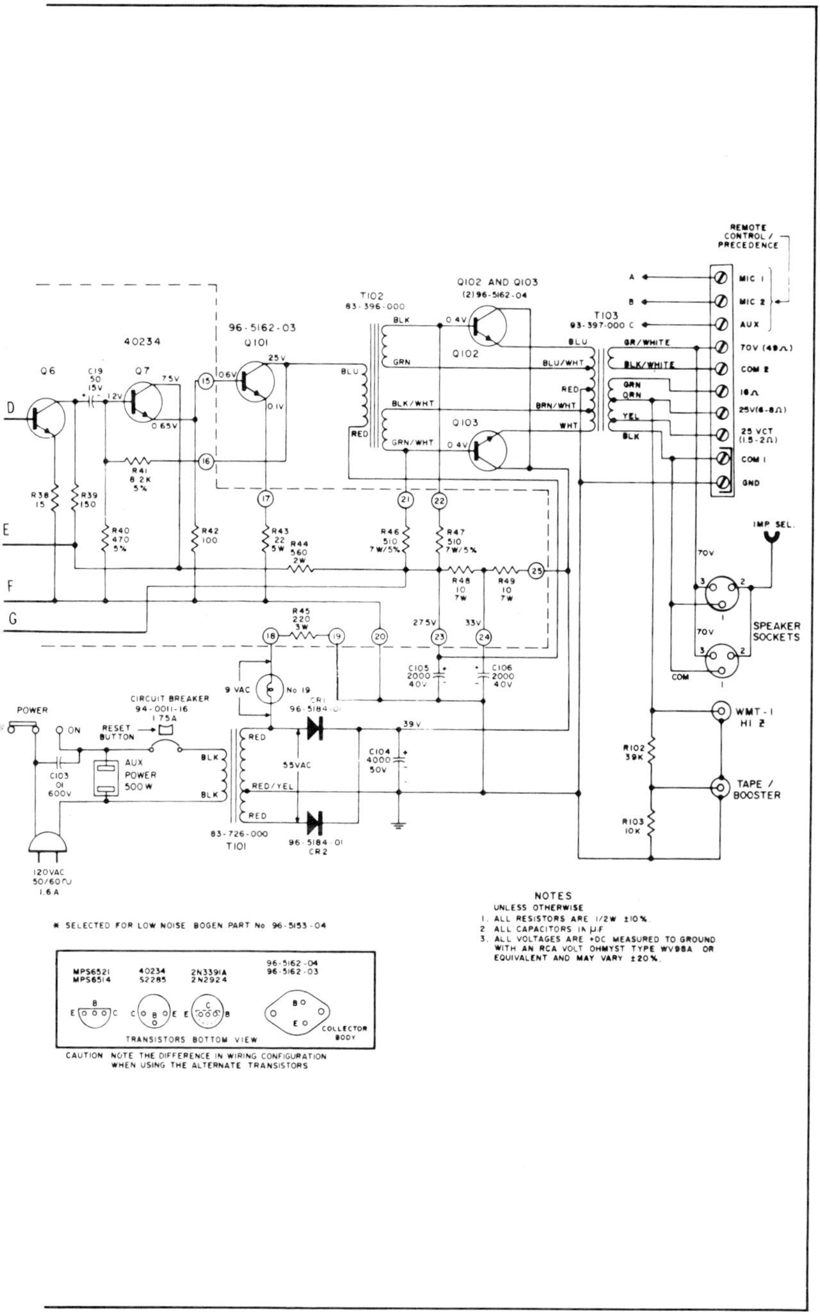

CHS50 amplifier.

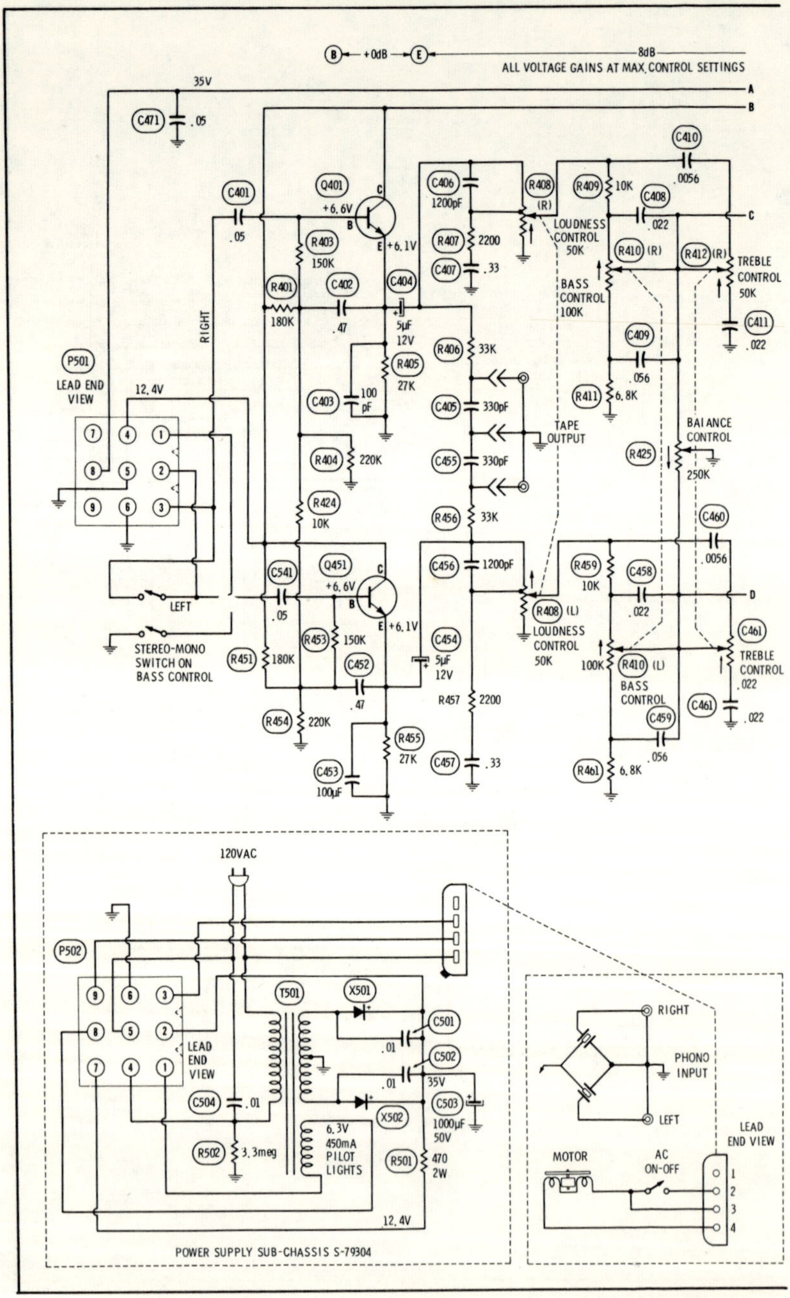

Fig. 6-10. Zenith

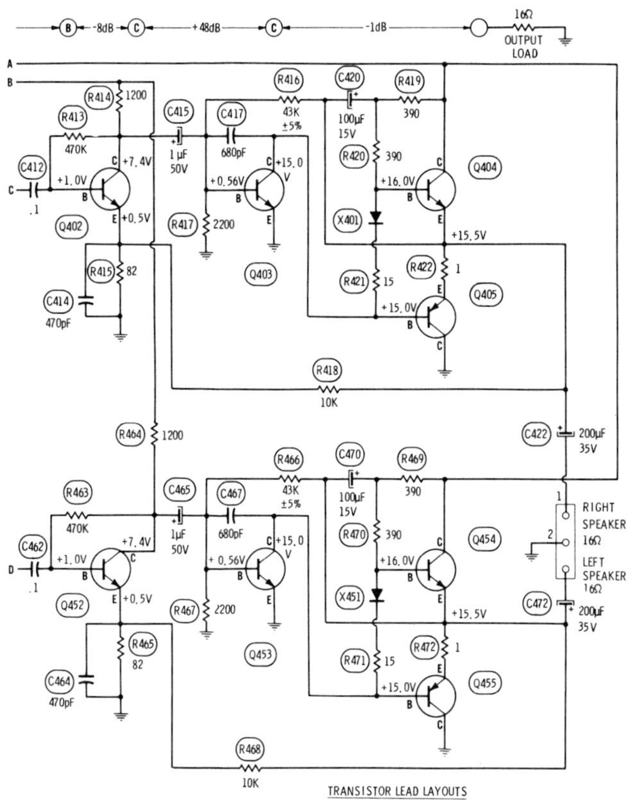

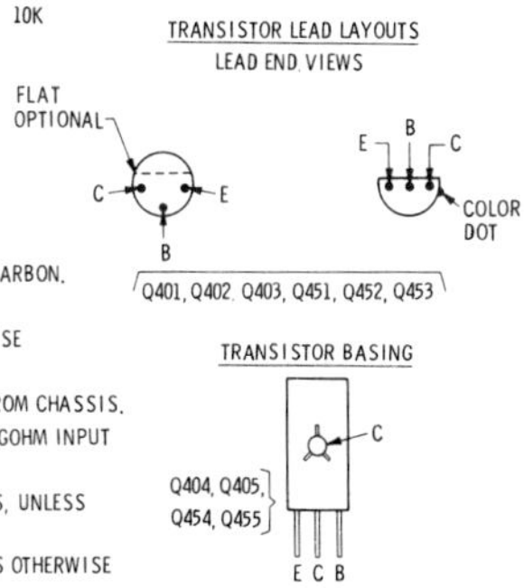

NOTES:

1. ⏚ INDICATES CHASSIS GROUND.
2. ARROWS ON CONTROLS INDICATE CLOCKWISE POSITION.
3. ALL RESISTANCES IN OHMS, 1/2 WATT CARBON, ±10% UNLESS OTHERWISE SPECIFIED.
4. ALL VOLTAGES ARE DC UNLESS OTHERWISE SPECIFIED.
5. DC VOLTAGES SHOWN ARE MEASURED FROM CHASSIS, WITH NO SIGNAL USING VTVM OF 11 MEGOHM INPUT RESISTANCE, LINE VOLTAGE 120VAC.
6. ALL CAPACITORS ARE IN MICROFARADS, UNLESS OTHERWISE SPECIFIED.
7. ALL CAPACITY TOLERANCE ±10% UNLESS OTHERWISE SPECIFIED.

TRANSISTORS		
NO.	PART NO.	DESCRIPTION
Q401	121-433	PREAMPLIFIER
Q402	121-430	PREDRIVER
Q403	121-706	DRIVER
Q404	121-710	OUTPUT
Q405	121-709	OUTPUT
Q451	121-433	PREAMPLIFIER
Q452	121-430	PREDRIVER
Q453	121-706	DRIVER
Q454	121-710	OUTPUT
Q455	121-709	OUTPUT

10ZT31 amplifier.

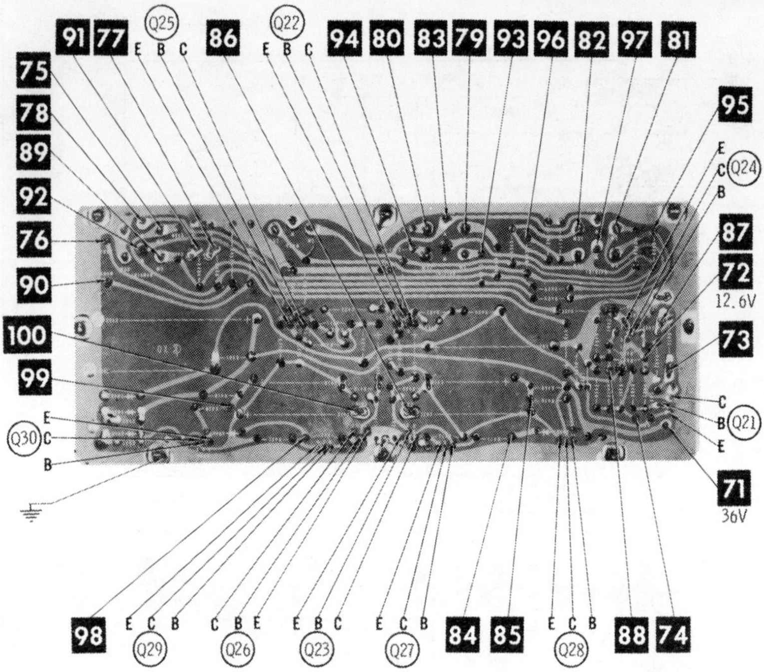

Fig. 6-11. PC board with CircuiTrace.

that the location of all transistors is shown by "callout" lines, and the elements of each one are distinctly identified. The *CircuitTrace* blocks match equivalent points on the PHOTOFACT schematic, and are a very big help in locating them; these are key test points in each important circuit.

In later model Sams schematics, the various part numbers will be the same as those shown on the PC board. This makes parts location even easier. Photographic illustrations of not only the underside of the board, but the top as well, make parts location possible.

SPEAKER SELECTION OR REPLACEMENT

Now that we're getting up into the area of higher power output, we are going to have to be especially careful when making speaker replacements not to inadvertently change the speaker load impedance! Vacuum-tube power output circuits are very forgiving of mismatches; you can do almost anything with them, and the only result is a slight loss of power, or some distortion. You can make horrible mismatches and get away with it without doing any damage to the amplifier.

You can *not*—repeat, *not*—do this in transistor circuits. Use speakers with exactly the same impedance as the original or you get into deep, deep trouble in only a fraction of a second. In pa-type output circuits with output transformers, you can get away with minor mismatches unless the amplifier is driven continuously to full output (which is not at all uncommon in musical instrument amplifiers).

Here's why. Every one of these amplifiers is *designed* to deliver a certain amount of maximum power output into a given load. The collector current, power dissipation in the transistors, etc., in OTL amplifiers, are most influenced by the load into which the two transistors work. In other words, this amplifier was designed for a certain power output with a certain load impedance, and any misguided attempt to change this causes trouble.

For an example, if you had an amplifier with a 50-watt power output and a 16-ohm load impedance, you would have a current of about 1.75 amperes through the speaker. (From here on, the actual figures are going to get "rounded off" most unmercifully. However, all I want to do is make the point, and they'll do that.) Fig. 6-12 shows this in normal operation. As you can see, this is a plain Ohm's law circuit, and the power in it will be determined by the current and resistance. This is the old familiar Watts $= I^2R$ equation.

Here, for normal output, 50 watts $= I^2 \times 16$ ohms, which gives us $50/16 = I^2$, and by the time you get through, you come up with about 1.75, which is close enough. Our driving voltage from the amplifier will figure out at about 30 volts, from $W = EI$ or $50 = 1.75$ E. Now, we have a voltage of 30 volts, driving a current of 1.75 amperes through a 16-ohm load. These conditions will cause a given amount of power to be developed in the load, and a fraction of that will be lost in each of the transistors. This is the "device dissipation" we talked about a while back.

Now, what would happen if we took off the 16-ohm speaker and replaced it with one of only 8 ohms? We would have practically

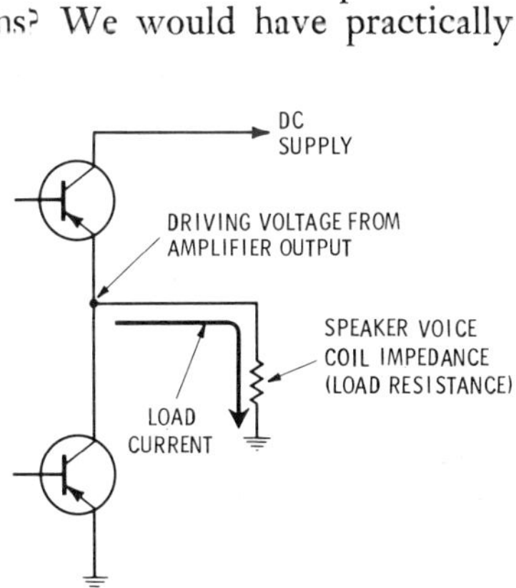

Fig. 6-12. Load circuit of OTL amplifier, showing how power is developed.

the same driving voltage, but what would happen to our power dissipation figures if the load resistance were cut in half? The same thing that happens in all circuits where Ohm's law applies (which is all of them). The *current* will double, since the resistance has been halved. As for the power output, power actually developed across the load by the driving voltage and current, this would change. It would go from the normal $1.75^2 \times 16 = 50$ watts, to $3.5^2 \times 8$, or 12×8, and you can figure that out for yourself as a resounding *96 watts.* This is practically *double* the amount of power developed normally.

Lest any of you get the optimistic idea that you *can* double the power of an amplifier by simply halving the load, let me remind you that if they had wanted you to get 100 watts out of this amplifier, they would have *designed* it for that much power! So, what you'd be doing in this case would be placing exactly a 100 percent overload on the output transistors and the power supply, and that won't work. Not for long, anyhow; at a guess, I'd say something like 100 milliseconds, and then something would go "phoop." The output transistors, probably. This would be "getting something for nothing," and nothing in electronics works that way.

So, if you get any questions about changing speaker impedances, the answer is a firm no. The only way you can make changes is by "rematching"; using two 8-ohm speakers in series to replace a 16-ohm, or two 16-ohm speakers in parallel to replace one 8-ohm. That's all. If you run into the problem of adding external speakers to a hi-fi system, for example, be very sure that you put back the same impedance you had to begin with. The only practical way of doing this is to switch the internal speakers off, and the outside or extension speakers on. Be sure that the selector switch you use for this is of the *nonshorting* type. Remember, you can *open* the load circuit of an OTL amplifier without hurting anything, but you cannot short it. (This is an extreme case of reducing the load impedance; it goes to zero, and Bang!)

Class-A amplifiers with output transformers, of course, are the opposite. They can stand shorts, but not open circuits. The same impedance-matching precautions still apply. Speaker-selector switches must be of the shorting type, or "make-before-break" to prevent the load from being open-circuited.

One sure method, with all systems and types, is to turn the power off before making any changes in the load. We have set up a few of these systems with the ac power switch interlocked with the speaker-selector switch. The switching sequence then is "INTERNAL SPEAKERS-OFF-EXTERNAL SPEAKERS." The ac power is turned off between the two positions. This is not at all hard to do, though it does require a few extra contacts on the speaker selector switch.

CHAPTER 7

Medium- to High-Power Types

Frankly, as an old tube-type technician, I never thought I'd see the day when I'd refer to a 100-watt audio amplifier as a "medium-power" type! They are now. Of course, an honest 100 watts of audio power is a whole lot. It can break windows if turned loose indoors without taking precautions. However, compared to some that you'll see in the following chapter, they are medium-power. Due to the differences in power supply needed, it's quite a lot easier to develop really high power with solid-state devices than it is with vacuum tubes.

Although, as we've said, power transistors do run hot, they don't develop nearly the heat that tubes do when you get up into the high power ranges. For one thing, in a tube amplifier of this wattage, the power supply rectifier tubes alone will put out more waste heat than the whole amplifier in a transistor type.

Fig. 7-1 shows the schematic of a 100-watt public address amplifier, a Bogen CHS-100. Here is more verification of the claim that these amplifiers are a lot alike. You can almost get a perfect match by putting the schematic of this amplifier over that of the CHS50 or CHS35. Only the dc voltage and current and the transistor types used are different.

The power transistors used here are special types. However, by checking collector voltage breakdown ratings, maximum currents, and power output, you should be able to find suitable replacements without too much trouble. The whole selection procedure will be explained in Chapter 10. TO-3 case transistors are the most commonly used. The major difference is that you'll see bigger and bigger heat sinks used on these more powerful amplifiers. In fact, this can be a sort of rough indication of the power output. The bigger the heat sink and the more fins, the greater the power it can dissipate.

Fig. 7-2 shows a solid-state "100-watt Booster" amplifier, Bogen MT-100. These are built to be driven by separate preamplifier units to get full output. However, only a 1.0-volt signal on the input will drive the MT-100 to full output. Both high and low impedance inputs are provided, so that any kind of input device can be properly matched. These units can be paralleled to get whatever total power is needed. In some instances, as many as five of these, totalling 500 watts of audio power, have been driven from a single preamplifier.

For the first time, you can see paralleled power transistors used in a standard output stage. To get the very high load currents needed,

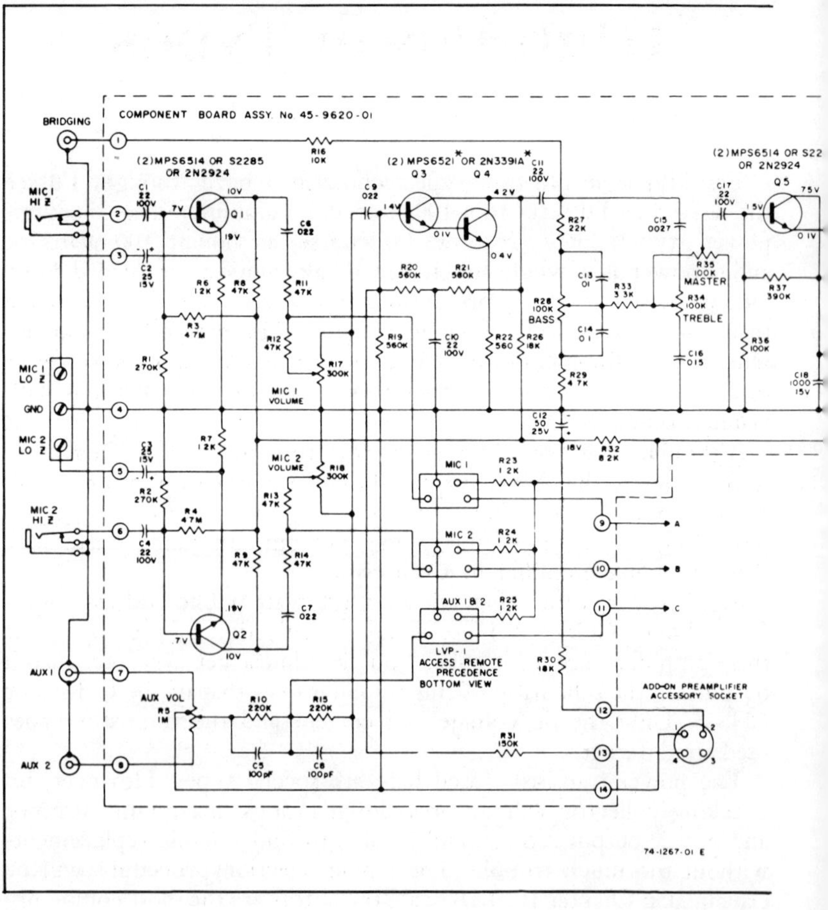

Fig. 7-1. Bogen Model

identical transistors are used. These must be "matched" for as close a set of characteristics as possible. Note the fuses provided in case one of the output transistors should short.

Drive for the output stage is furnished by a pair of direct-coupled transistors, Q7 and Q8, which are in turn driven by another pair, Q4 and Q5. The preamplifier stage is also direct coupled. In fact, this three-transistor circuit is called a "Darlington Triplet": Q1, Q2, and Q3. Notice that Q1 and Q2 form a standard Darlington pair, and the collector of Q2 is directly coupled to the base of Q3. The dashed lines here mean that this unit is built on a small separate PC board, not that these three transistors are in the same case, just as the Darlington types were in the smaller amplifiers you have studied. This triplet circuit has extremely high gain; it works just like the Darling-

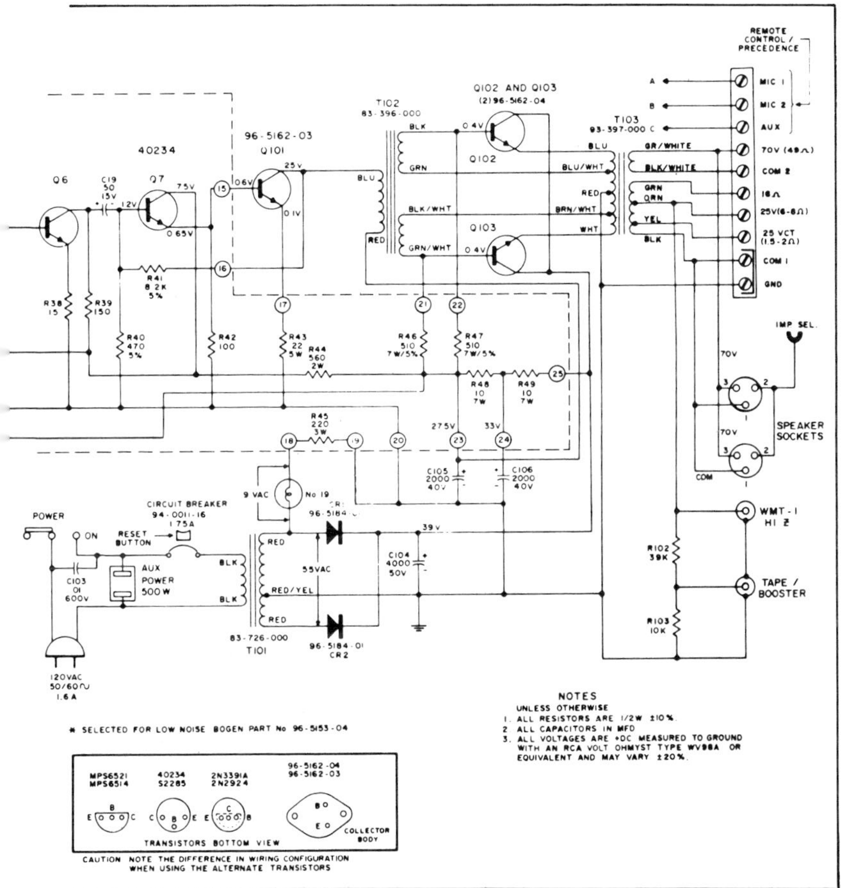

Courtesy Bogen Communications Division, Lear Siegler, Inc.

CHS100 PA amplifier.

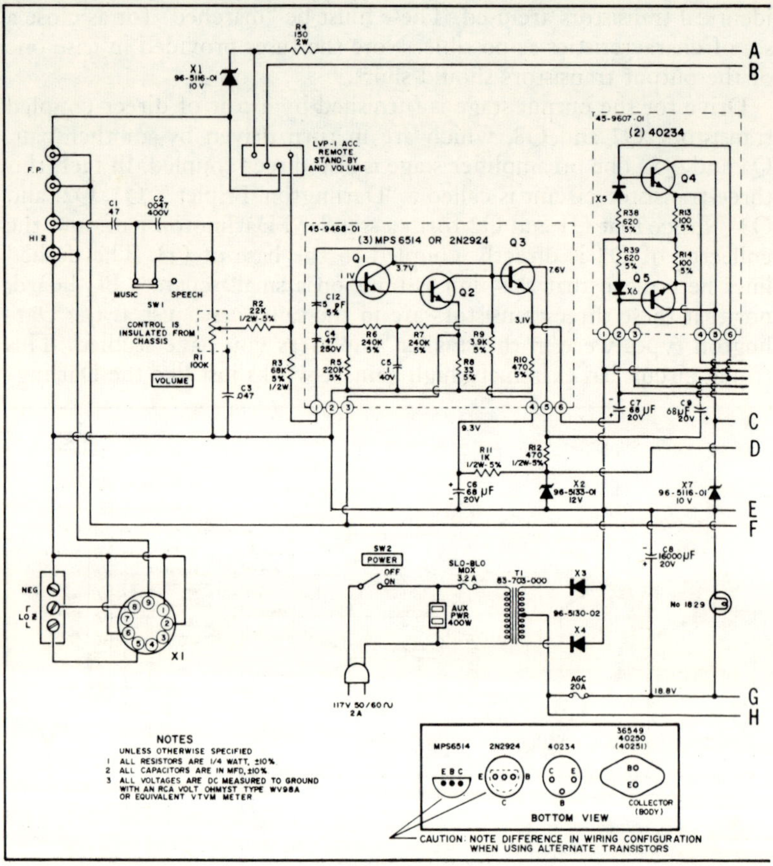

Fig. 7-2. Bogen MT100

ton pair you have seen, with an emitter-follower driving a common-emitter, which in turn drives another common-emitter stage in cascade.

If one or more of the power-output transistors burn out, it would probably be best to replace them all. For best results, and especially for the right thermal characteristics, these transistors must be matched as closely as possible. When one goes out, the resulting surge will probably do some damage to the rest. So, it's better to avoid the chance of possible future failure by replacing them all.

160-WATT HOME STEREO AMPLIFIER SYSTEM

Fig. 7-3 shows a 160-watt, 80-watts-per-channel, stereo amplifier, a Zenith 6YT24. Once again, we see the same basic circuits in use—

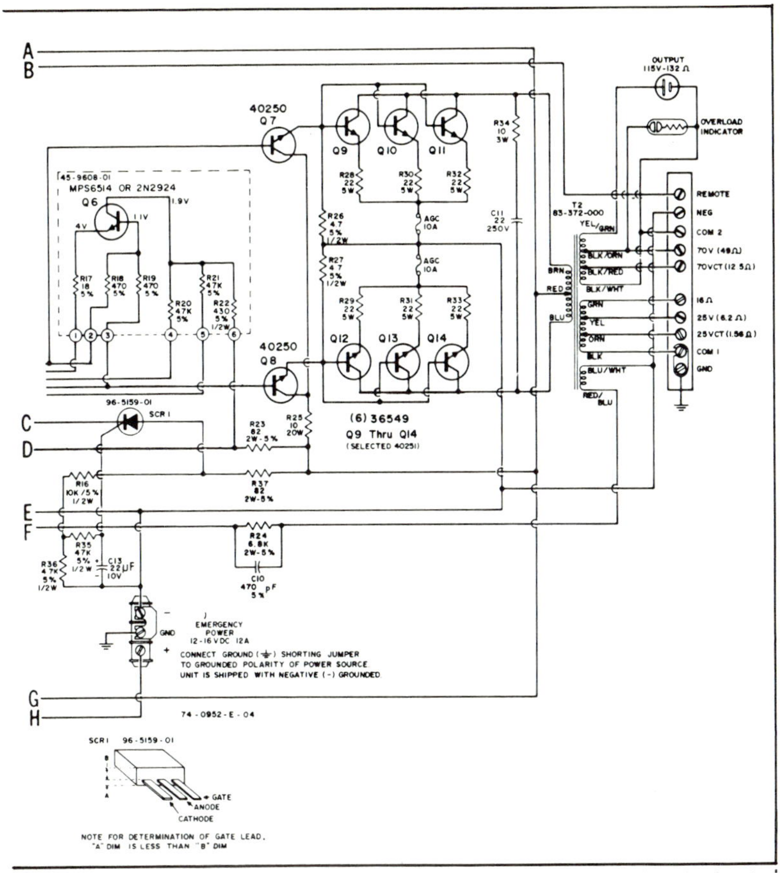

Courtesy Bogen Communications Division, Lear Siegler, Inc.

"booster" amplifier.

stacked-pair output transistors, an input transformer, and a driver transistor. The preamplifier stage for this unit is on the phono or radio chassis used with it. Fig. 7-4 shows the layout of the audio-amplifier chassis. Note the size of the heat sinks used for each output pair. Also, note that now the *driver* transistors are big enough to need heat sinks too. This is done by mounting them on the end of the amplifier chassis. All of these are TO-3 cased type transistors.

HEAVY-DUTY RECTIFIERS

Now, we see something else that's new. Note the silicon diode rectifiers used. Up to this time, all we've seen are the small axial-lead packages, like the ones used in tv sets. These can handle currents up to 1.0 ampere without trouble. Now we're getting into

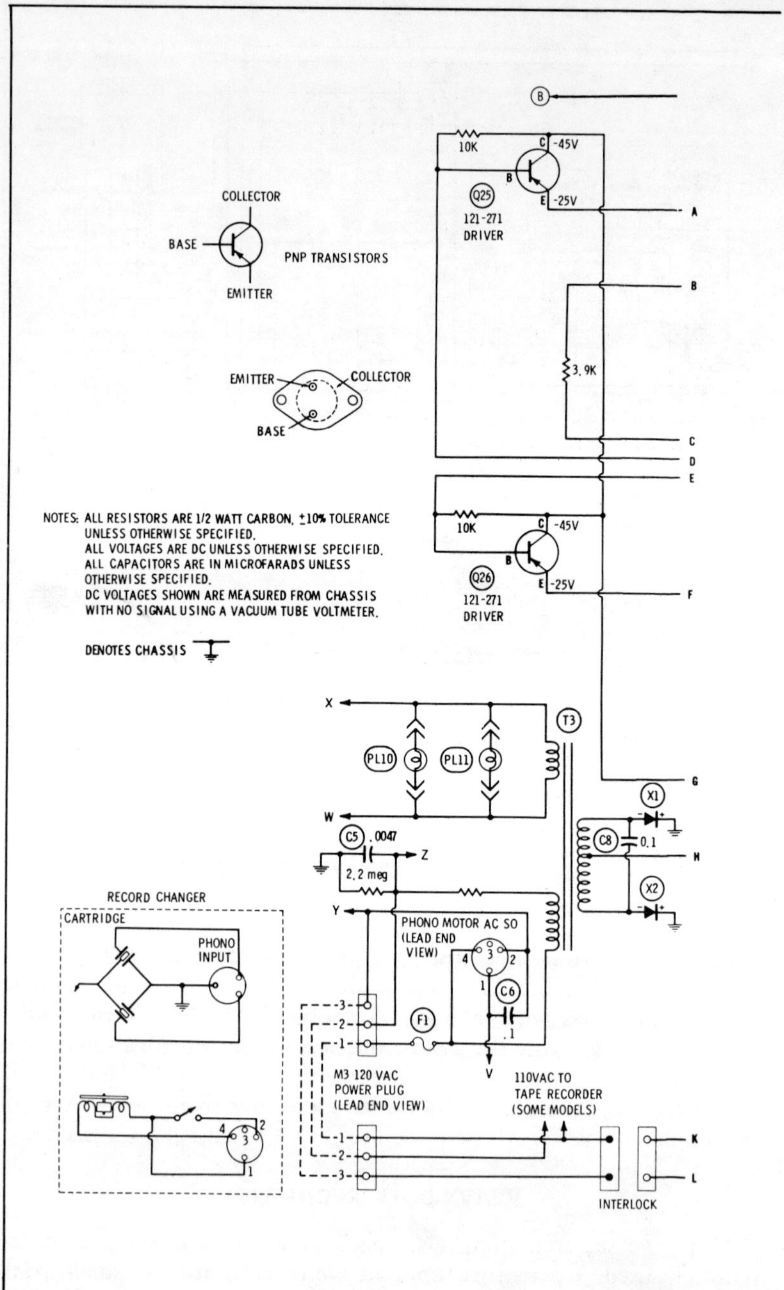

Fig. 7-3. Zenith

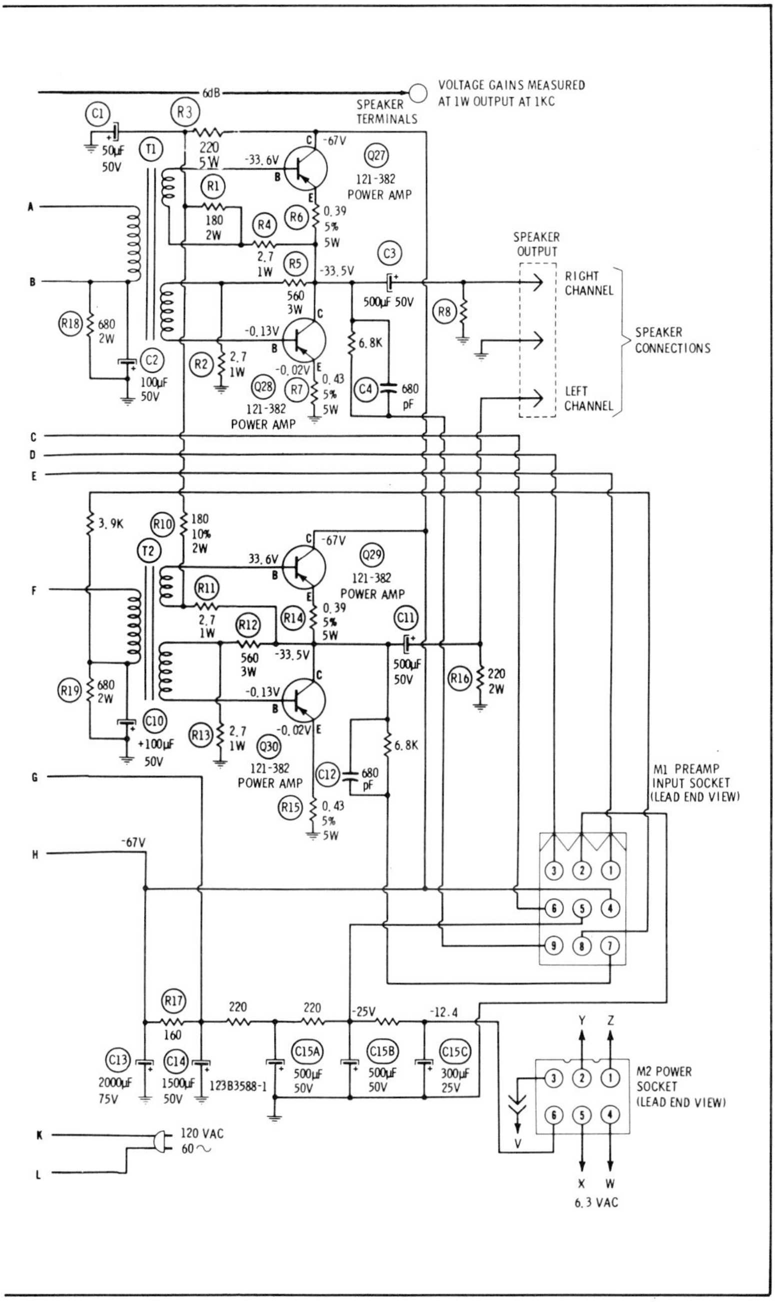

6YT24 stereo amplifier.

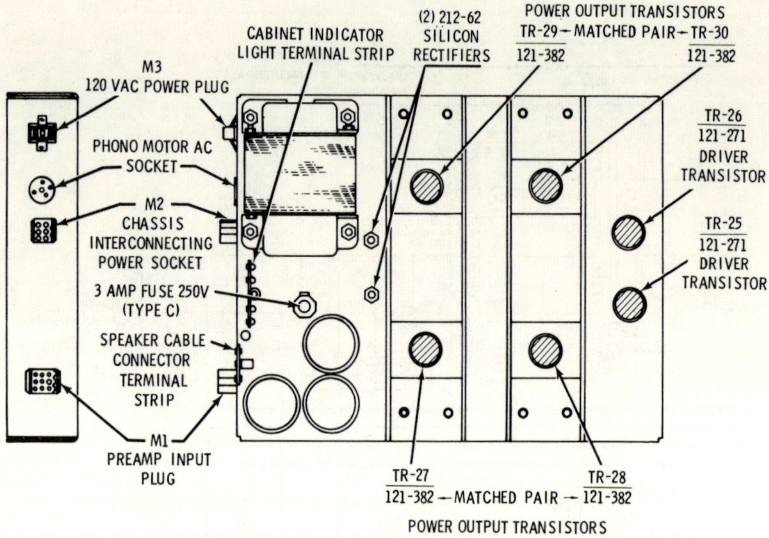

Fig. 7-4. Chassis layout of Zenith 6YT24 amplifier.

much higher current drains; so, the rectifiers must also have heat sinks to keep them cool.

The diodes used here are called "stud-mounted." Fig. 7-5 shows this type of mounting. The correct name for this is a DO-5 case. The threaded stud is common to the case, electrically and, in the original package, was connected to the cathode of the diode. The anode is connected to the insulated terminal on the top. If the case and stud cannot be electrically connected to the heat sink, they can be insulated with the mica and Teflon washers shown.

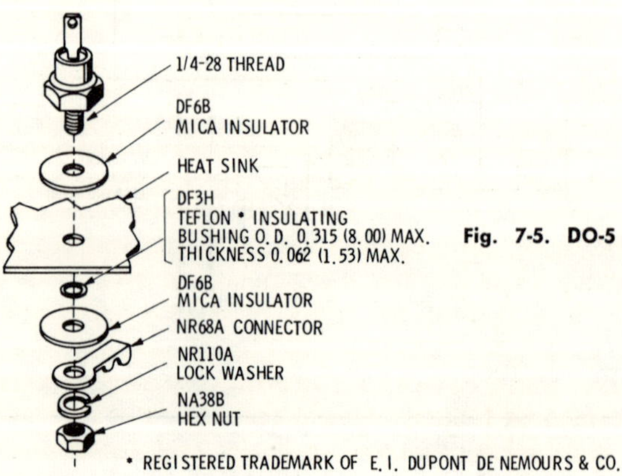

Fig. 7-5. DO-5 stud-mounted heavy-duty rectifier.

Replacements for these are available in several lines. The Motorola HEP-153 is an electrical duplicate, having a 200-volt piv rating, and a current maximum of 15.0 amperes. The RCA SK-3500 is also the same, with a 600-volt piv and a current maximum of 12.0 amperes.

In the Zenith amplifier power supply, the schematic will show you that both cathodes are grounded. So, the diodes are bolted directly to the chassis; the anodes go to the ends of the power transformer secondary, and the negative dc voltage is taken from the center-tap of the winding. This is the standard full-wave rectifier circuit.

You will also run into cases where these diodes have the connections *reversed*. The anode is connected to the stud, and the cathode is connected to the insulated terminal. These are usually denoted by a letter "R" at the end of the type number, meaning "reversed." Watch out for this when ordering replacement rectifiers.

65 AND 125-WATT AMPLIFIERS

As another example of the similarity of these circuits, the General Electric T65A and T125A amplifiers are practically identical. Actually, only seven parts are changed to make the basic amplifier from a 65-watt output into a 125-watt output: two resistors (bias), the four output transistors and the power transformer. Fig. 7-6 shows the T125A circuit.

Once again, we see the dual-polarity power-supply voltage, taken from a full-wave bridge rectifier, with a center-tapped secondary winding on the power transformers. This means no blocking capacitors between the output pair midpoint and the speakers, and zero dc voltage on the midpoint. One quick check for circuit identification between this type and the standard stack or comp-symm pair, is to read the voltages on the "outside elements" of the output transistor pair. Here, this would be the emitter of Q20, +33 volts, and the collector of Q21, −33 volts. In the single-voltage power supply type, one would be a dc voltage, the other ground. So, if you find zero voltage at the midpoint, check the other end of the circuit, and you can tell which type you have. (Of course, this is for circuit identification when you don't have the schematic. If you have one, as you should, it'll be easy.) The electrolytic capacitor you see in there is not the blocking capacitor but a nonpolarized type with a value of only 10 μF; it is part of the speaker crossover network.

Three speakers are used for each channel. A 12-inch woofer (bass) and a 3-inch "midrange" plus a tweeter of the horn type. Net speaker load on this output stage is 8 ohms. A selector switch allows the user to select either the internal speakers or an external (extension)

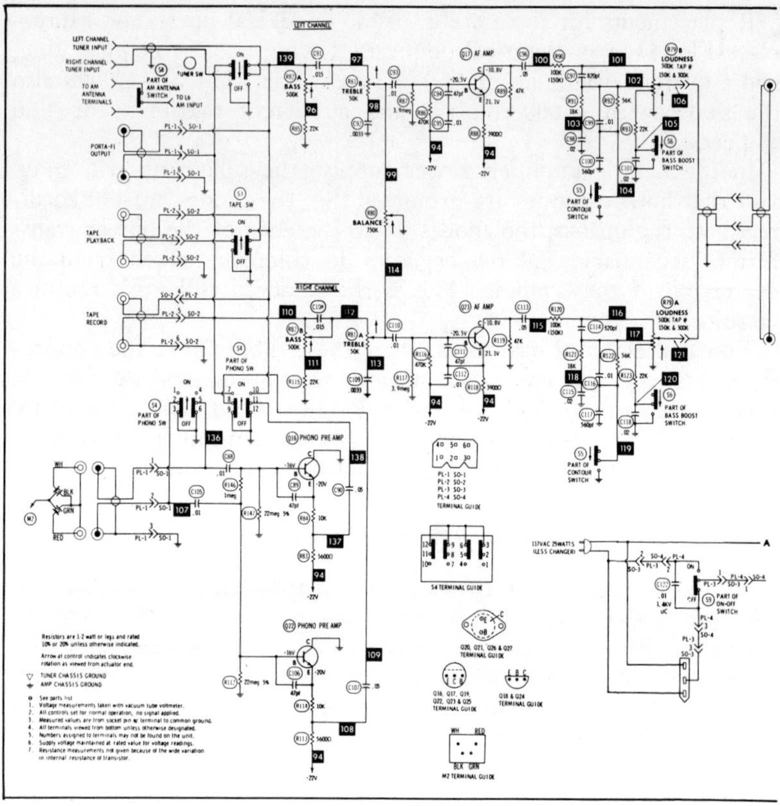

Fig. 7-6. General

speaker system. A shorting link, seen above the selector switch, will allow both channel outputs to be fed into a single extension speaker for patio or poolside use, etc.

A full-wave bridge rectifier, with 1.0 ampere peak type "top-hat" silicon diode rectifiers, provides all dc voltages. Maximum current drawn from the power supply at full volume will be about 1.0 ampere. However, with a bridge rectifier there are always at least two rectifiers carrying the current. So, these are not really overloaded, as it might seem.

Aside from this, these amplifiers are practically the same as the rest. A two-stage, direct-coupled driver feeds the input transformer, to develop a large enough drive signal for the output stage. Bias-diode clamps are used on each of the high-power output transistors. These are germanium types, and more prone to thermal runaway than the silicons. The diodes are mounted on the same heat sinks; they will stay at the same temperature as the output transistors. If

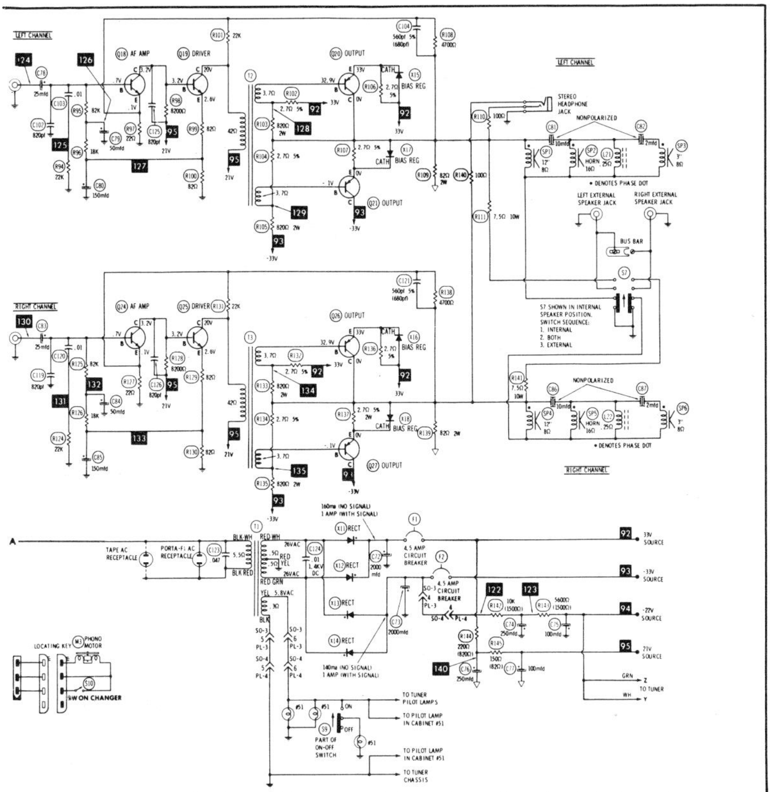

Electric T125A amplifier.

the transistors start to heat up, the bias diodes do, too. This causes a change in their voltage drop, in the reverse bias direction, which pulls the collector current down, and cools things off. Incidentally, whenever you see these protective diodes used in a circuit, they *must* be of the same type as the output transistors, germanium or silicon, in order to get the correct thermal reaction. Watch out for this when making replacements of any diodes which have shorted or developed other defects.

POWER-OUTPUT TESTING

When we make performance tests on these amplifiers, we need to know two things to be able to tell whether the amplifier is working normally. One is the total power output the amplifier can deliver without distortion, and the other one is the sensitivity. Meaning, "How much input signal is needed to drive this amplifier to full power output?" Let's take power output first, since it provides us with most of the trouble.

There is a standard test procedure used by many manufacturers. It's quite simple, with only a few pieces of equipment needed. All you need is an audio signal generator, a scope, an ac vtvm, and a great big resistor. To make the test hookup, disconnect the speakers, and connect the load resistor in their place. This load resistor must be exactly the same value as the amplifier output impedance; check this on the schematic. Also, and just as important, it must have a high enough wattage rating to handle the full output of the amplifier. In the case of these General Electric amplifiers, this would mean a minimum of 65 watts for one channel of the T125A, since we can check one channel at a time. For safety, we should use at least a 75-watt type, and a 100-watt wouldn't be too bad. You'll need it at some later date anyhow.

It's possible to get large resistors with an adjustable slider. The slider is held by a clamp screw, so it can be set for whatever resistance value is needed. Remember this fact, though; if you have, for example, a 100-watt, 16-ohm resistor, and you set the clamp for the center for an 8-ohm load, you *now* have an 8-ohm resistor at 50 watts, not 100. (Wattage dissipation rating is for the whole resistor.) If you have an "even split" like this, you can get your 100-watt rating back by centering the slider and then tying the two ends of the resistor together. Watch your total resistance, though; here, you'd have two 8-ohm resistors in parallel, and only 4 ohms resultant. You'd need a 32-ohm resistor, which would give you two 16-ohm resistors in parallel, for a resultant of 8 ohms. Plain Ohm's law. You can build up paralleled resistors of the right values to get any resistance and wattage values needed. For example, four 64-ohm resistors, at 25 watts, in parallel, and you come out with one 16-ohm at 100 watts, and so on. Fig. 7-7 shows the "halved" resistor setup. (You can often get very high-wattage wirewound resistors from surplus at a low cost.)

Now, we're finally ready for the test. Fig. 7-8 shows the equipment setup. The audio signal generator is connected to the phono input of the amplifier. Set this to 1000 Hz; we used to make audio tests at 400 Hz, but the 1000 Hz frequency is now more nearly a standard. (When you find frequency response characteristic curves, note that they are almost always set up with the 1000 Hz response as zero dB or the reference point. A great many audio test records also use this frequency.) The scope and ac vtvm are connected directly across the load resistor as shown. Of course, you could read the peak-to-peak voltage on the scope, but then you would have to convert this to rms values for the computation of power, and it is actually faster to use the ac vtvm, reading rms values.

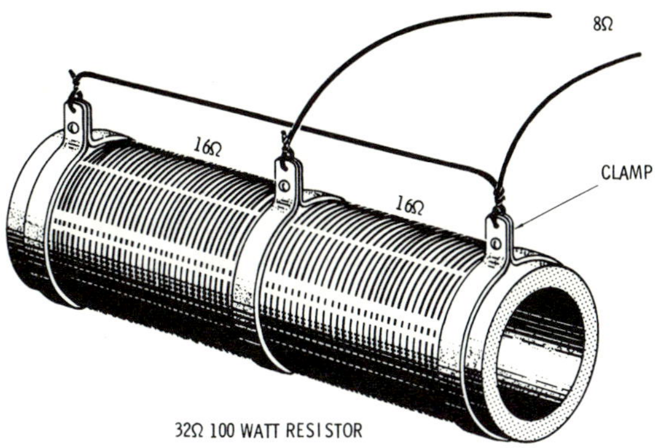

Fig. 7-7. Heavy-duty load resistor.

Now, the loudness control is turned to maximum, and the tone controls are set at center, in most cases. In some specs you find that these are set to full cw, etc. Any tone control switching, such as special bass boost, etc., is turned off.

Turn the power on, and gradually raise the input signal level until you see a sine-wave pattern on the scope. Keep on increasing the signal input, reducing the scope vertical gain to keep the pattern on the screen. Continue increasing the signal until you see the peaks of the sine wave on the scope just barely begin to show signs of flattening. (Actual peak-to-peak voltage shown on the scope is not important; what we're using it for, here, is merely to display the waveform.) When you see this, stop, and back down the signal input until the clipping of peaks has just barely disappeared. This is the full undistorted power output of this amplifier, equivalent to the

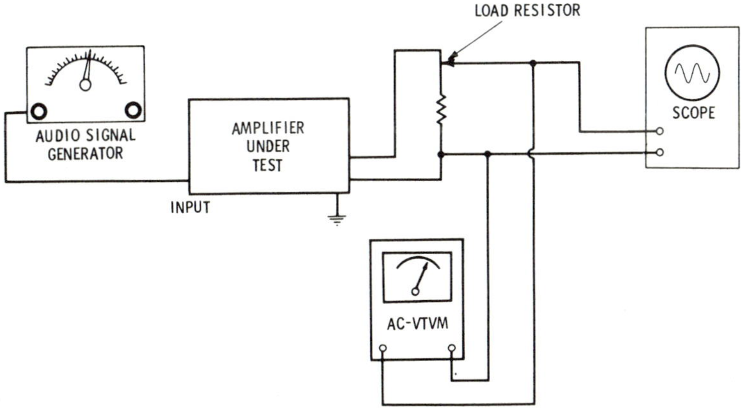

Fig. 7-8. Test-equipment setup for power-output tests.

IHF/EIA rating of 5 percent third-harmonic distortion, or music power.

Now, using the vtvm, measure the rms voltage across the load resistor. For the T125A amplifier, this should be about 27 volts for the full 65 watt-per-channel output. The factor specs show a *minimum* voltage of 18 volts rms as an acceptable level, which corresponds to an output of about 45 watts. You can check this for any amplifier, by using the Ohm's law E^2/R equation. Square the output voltage and divide it by the load resistance, and there's your rms power output.

SENSITIVITY TESTING

In the preceding test, we actually checked only the power output capability of the amplifier. In other words, *can* it be driven to produce its full rated power output? This checks mostly the power-output stages. Now, we really need one more test to make things complete, a voltage-sensitivity reading. In other words, *how much input signal* do we need to drive this amplifier to its full power? This measures the signal-voltage gain of the preamplifier stages, drivers, etc.

You can use exactly the same test setup as before. You don't really need the ac vtvm for this, on the output. Turn the audio signal generator output up until you find the maximum undistorted level on the output. Now, read the ac signal voltage across the *input*. This can be done by reading the setting of the calibrated attenuator, if your af signal generator is equipped with one, or by reading the signal voltage directly with the ac vtvm.

In the past, some manufacturers used a 50-millivolt input signal as a standard. Now, a 400-millivolt input seems to be more popular. This is the case in the sensitivity tests given in the factory service data for the two General Electric amplifiers. The 400 millivolts is a level that can be used on any amplifier, even if you don't happen to know the factory level for tests. If any amplifier will go to full output at this level of input, it should be all right. The typical input signal, from phono pickups and guitar pickups, etc., should be at least that much, and more in some high-output types.

If the amplifier will go to full output at a level very close to 400 mV, it is in good condition. If you find small discrepancies, check the ac line voltage; these tests are usually specified as being made at a 120-volt ac input to the amplifier. Serious deviations from this normal input level will show you that there is trouble somewhere in the voltage-amplifier or "small-signal" stages of the circuit. Stage-by-stage gain checks will help you find out where the loss is located. (In quite a few instances, you may find open transistors causing low gain. Because of the very close spacing between transistor elements, the signal will often feed through, instead of "stopping dead" as a dead tube will do.) If the loss of gain is accompanied by a pretty severe increase in distortion, especially clipping, look for a leaky preamp or driver transistor. This upsets the bias, causes clipping, and reduces the gain of the stage. The scope will locate such troubles in signal-tracing tests through the amplifier.

If your audio signal generator doesn't have an accurately calibrated output attenuator or if you don't have an ac vtvm that will read down to 400 millivolts, don't worry. (However, 400 millivolts is 0.4 volt, and if your voltmeter has an 0-1 volt or 0-1.5 volt scale, it should read it very well, since this will be in the most accurate part of the meter scale, near the middle.) You can make up your own voltage divider; get a 600 and 400 ohm resistor, or a 1000-ohm resistor with a slider. Connect the audio signal generator across the whole resistor, as shown in Fig. 7-9.

Set the signal generator to 1000 Hz, then adjust its output to give a reading of 1.0 volt across the resistor. The tap will then be exactly 400 millivolts. This is 400/1000ths of one volt. Any other values needed can be picked off this setup if the resistor is adjustable. Although wirewound resistors are slightly inductive, they can be used for this test, since their inductance at 1 kHz is pretty small.

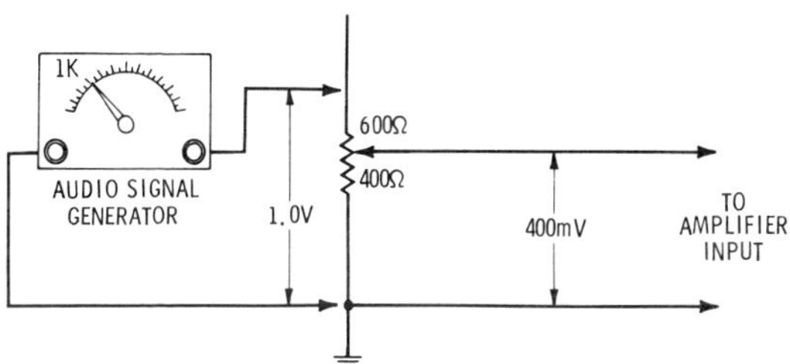

Fig. 7-9. Input voltage divider for reducing signal levels.

POWER RATING OF SPEAKERS

From this point on (we should actually back up far enough to take in the 35-to 50-watt amplifiers), we're going to have to consider one more very important thing when selecting replacement speakers. In addition to their size, shape, and frequency response, plus the voice-coil impedance, we must consider their *power rating*. Back in what some persons insist on calling the good old days, cone speakers had a maximum power rating of about 5 watts, and if we wanted more than that in a single speaker, we had to use the "driver and metal horn" pa-type outdoor speakers. If we had to handle 50 watts, the only way we could do it was to use two 25-watt drivers in parallel.

Not any more! The need for very high-power cone type loudspeakers started a few years ago, when the craze for higher and higher power in musical-instrument amplifiers began. Now, it is possible to get cone-type speakers with a power handling ability of 100 watts. Needless to say, these are not the thin cardboard cone, fine-wire voice-coil types that we used before. (On one occasion, a musician bought a very powerful amplifier and what the salesman assured him were high power speakers—four of them. The next day, he was back. When he had turned the super-powered amplifier wide open and hit a chord on the guitar, he had literally turned all four of the speaker cones inside-out! In two of them, the grille cloth had been blown out as well.)

Modern heavy-duty speakers of the required type are expensive, but they're the only ones that can take the punishment. Voice coils now are made of thick aluminum ribbon, cemented to the forms with epoxy resin which gets harder when it gets hotter. In some of the tests used in developing these speakers, voice coils of the old type were literally melted by the tremendous currents flowing through them. When the voice coils were made heavy enough so that they wouldn't burn in two, the cement softened and the voice coils were pulled off the forms. However, all of these problems have been solved, and today's speakers can really "stay in there and take it."

The worst disadvantage of this type of speaker is that it must be sent back to the factory for repairs when needed. The average reconer doesn't have the facilities for taking them apart and repairing them. The days of "soften up the cement with acetone and put in a new cone" are gone. Due to the high cost of this type speaker, it is much less expensive to re-cone it than buy a new one.

In the multiple-speaker systems used in stereo theater, and even in some musical instrument amplifiers, all speakers must be replaced with exact duplicates. In multiple systems, the replacement must have exactly the same impedance as the original, so as not to upset the impedance matching to the amplifier. Fig. 7-10 shows the typical speaker circuitry found in the large multiple-speaker systems. This is the speaker circuitry of the Zenith 6YT24 amplifier, as used in quite a few models.

Note the somewhat unusual combination of impedances of the four speakers, compared to the 8- or 16-ohm impedance of the single-speaker amplifiers we have seen before. This will "come out" at about 8 ohms or a little less. Others will differ slightly.

The bass speakers or woofers will probably be the most important factor in determining the power output. Due to the characteristics

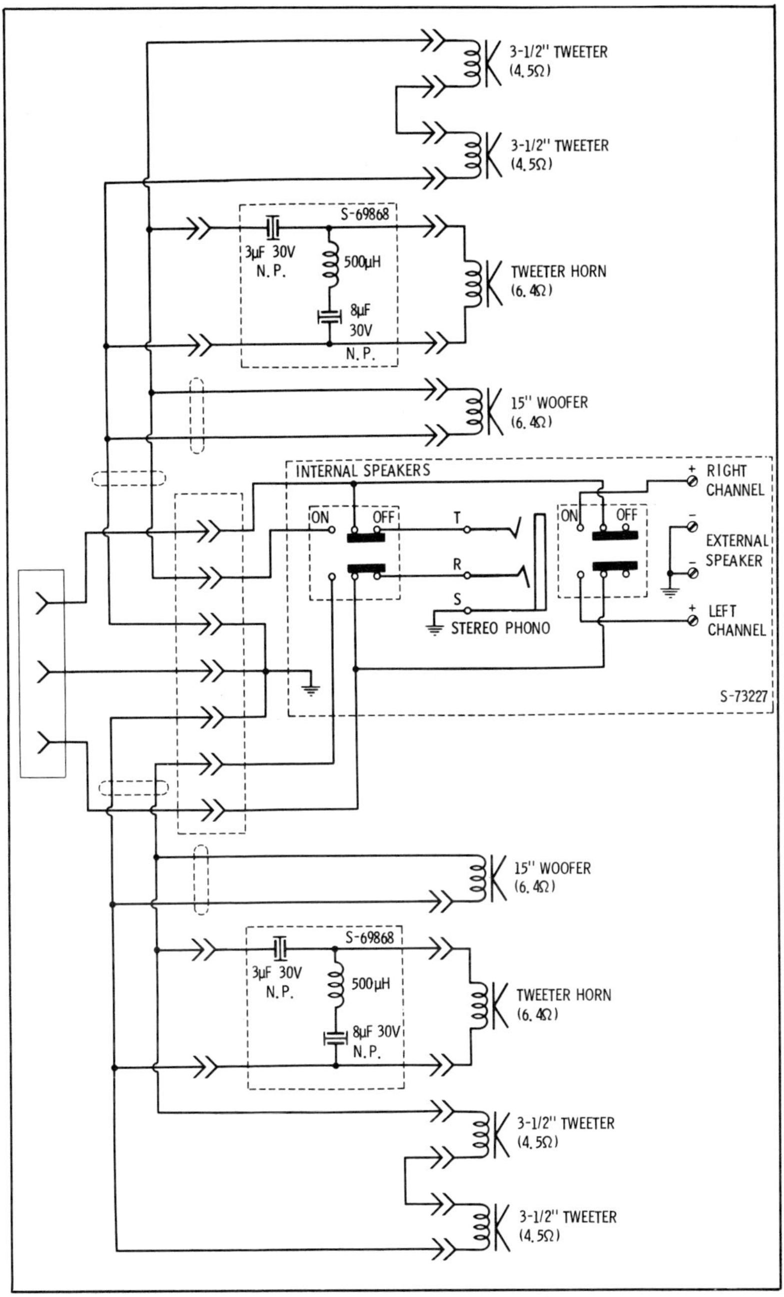

Fig. 7-10. Zenith 6YT24 speaker systems and switching.

of sound, something like 65 percent or more of the actual audio power in a composite signal (such as music) is in the low frequencies. The bass speakers have to do more work in moving air-masses, therefore they need more power. The tweeters will actually have to dissipate only a small part of the total audio power, because of their small horns or cones and the reduced motion of their diaphragms. (In some of the later model woofer speakers, the cones can move as much as 2 inches. This means that they can really "put out the power" in moving air, converting electrical power into acoustical power or air in motion.)

CHAPTER 8

Very High Power Amplifiers

Now, we come to the "Big Boomers"—the tremendously high-powered transistor amplifiers that have been developed in the past few years. The majority of these will be found used by "groups" for musical-instrument amplifiers; lead, rhythm, and bass guitars; and for amplifying vocal choruses. The largest of these will probably be in the class called "Bass Amps." These are used to amplify the bass guitar to provide a solid rhythm background for the whole band.

Some of these amplifiers have power outputs up to 300 watts. This is probably peak power, since most of them seem to be played "flat-out" most of the time. However, there are several super-power instruments made for use in the home. One of these, at least, has a music-power output of 250 watts. (Peak, 500 watts!) The average high-school football stadium can be well-covered with an amplifier of not more than 50 watts, so you can see why we call these "super-powered" types.

THE RCA RS-238A POWER AMPLIFIER

Fig. 8-1 shows the power amplifier circuit of the RCA RS-238A power amplifier. This is a stereo amplifier, with a peak power rating of 250 watts per channel and a music-power rating of 250 watts for both, 125 watts each. Once again, we see the same circuits used. Only the voltage and current ratings of the power transistors, rectifiers, and the power transformer used in the circuit have been changed. A dual-voltage power supply is used, with dc outputs of +36 and −36 volts.

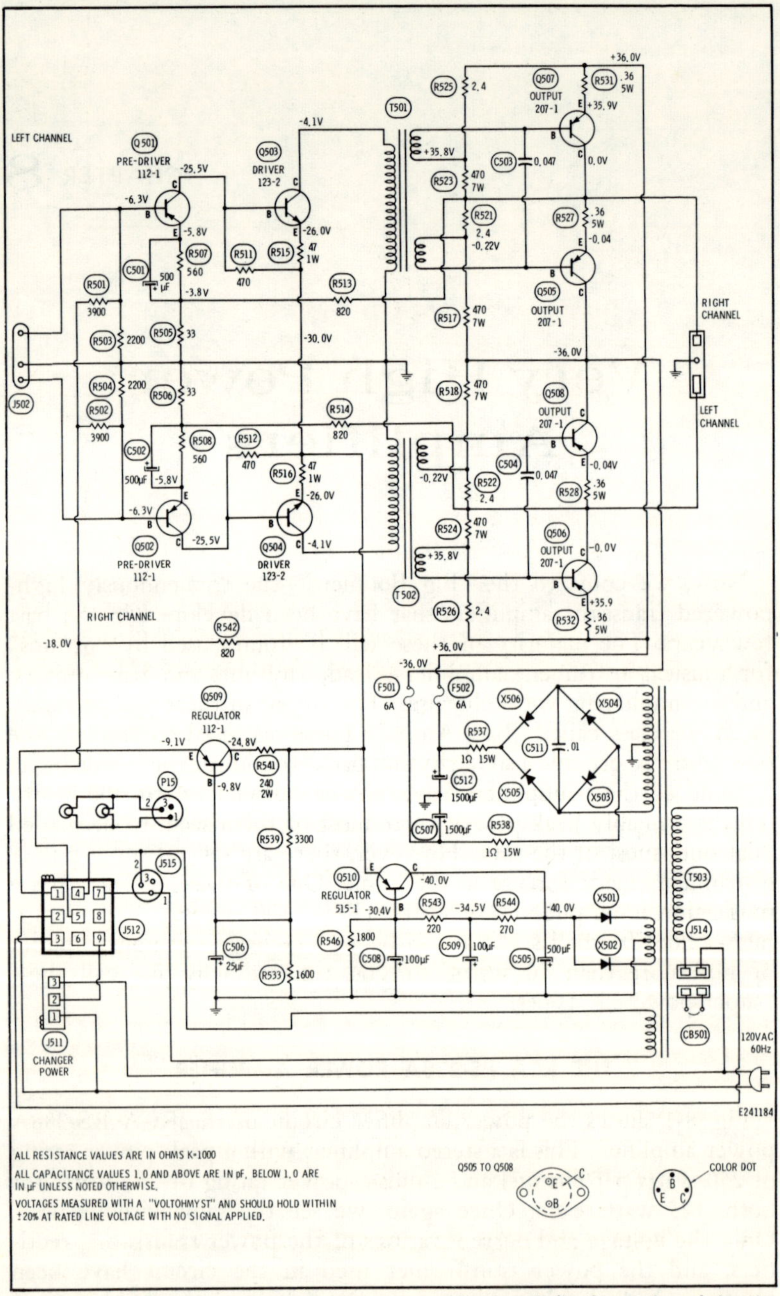

Courtesy Radio Corporation of America

Fig. 8-1. RCA RS-238A power amplifier.

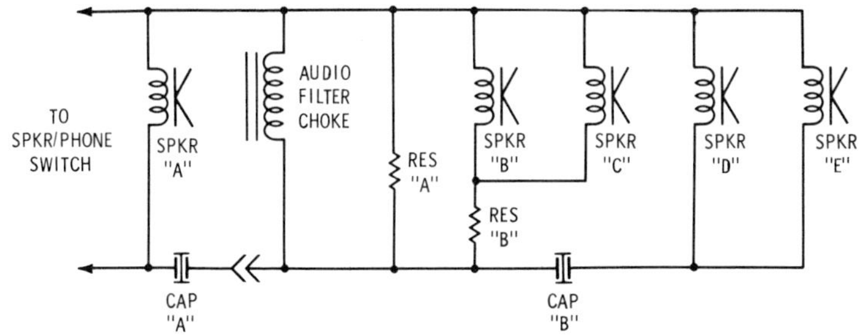

Fig. 8-2. Speaker wiring diagram for RCA RS-238A.

A total of ten speakers are used, 5 for each channel – a 15-inch woofer with an 8.5-ohm voice coil, two midrange horn speakers with 12-ohm voice coils (connected in parallel), and two 3.5-inch tweeters with 20-ohm voice coils, also in parallel. This combination comes out somewhere around 2.0 ohms. As you saw in the power-output calculations in a previous chapter, the lower the load impedance on an OTL circuit, the higher the current, and the greater the power output developed. As in any parallel circuit, most of the audio power will appear across the lowest impedance, which is the woofer with its 8.5-ohm voice coil. The others will divide up the remainder according to the ratio of their impedances. Fig. 8-2 shows the speaker hookup used.

All speakers are enclosed in tightly sealed compartments in the ends of the cabinet. This provides an "air-cushion" effect, which prevents the woofer cones from making too great an excursion under full power. This is something you'll have to be careful with. This type of high power speaker system should *never* be operated at full power, *without* the back of the speaker enclosure in place for a tight seal! If it is, the woofer cones lose their damping and can be driven so far that the suspension is damaged, or the cones can be torn loose from the rims. More on this later.

Fig. 8-3 shows the chassis layout of the RS-238A amplifier. Note that now, in the high-power type, even the driver transistors (Q503 and Q504) use their own heat sinks. The others with heat sinks are the voltage regulators. In this circuit, the power output is provided with its own separate dc supply, at +36 and −36 volts, unregulated, and the drivers have a regulated −30-volt supply of their own. Power for this comes from another secondary winding on the transformer.

BASS AMPLIFIER FOR MUSICAL INSTRUMENTS

Fig. 8-4 shows the circuit of a high powered "Bass-Amp" musical instrument amplifier. This is the Magnatone PS-300. Here we see a

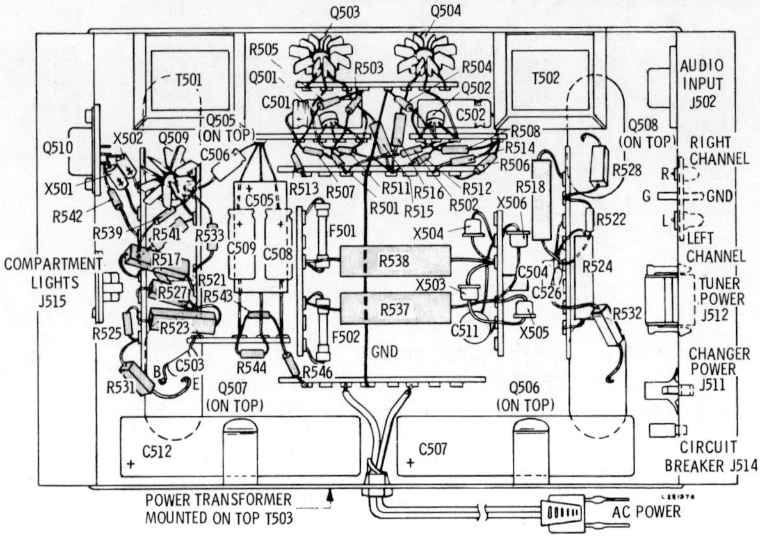

Courtesy Radio Corporation of America

Fig. 8-3. RS-238A power amplifier layout.

characteristic of the super-powered types, as of this writing; the amplifier output is divided into two channels, each having a separate high-power output stage driving a single 15-inch speaker. These are 4-ohm very heavy-duty types. Here, two small preamplifiers are used, with their outputs in parallel driving the two power amplifiers. So, the maximum output is available from any of the four input jacks.

In other versions of this instrument, more preamplifiers are used, or separate vibrato and reverberation circuits (each on its own small PC board), for special effects. A master volume control sets the gain of the power amplifiers; as usual, each musical instrument has its own volume control on the body of the instrument itself.

The standard quasi-complementary-symmetry output circuit is used. This circuit uses identical output transistors; the complementary-symmetry circuit is in the driver stage. The output transistors here are special types with in-house numbers, not often listed in replacement guides. However, they can be replaced by stock types, as we'll see in Chapter 10.

Elaborate filtering is used to hold down the hum. A brute-force filter is connected to the bridge-rectifier output. Note that the input capacitor is actually 8500 microfarads. Two 4250 μF capacitors are connected in parallel, rated at 75 volts, working. A three-section RC filter network with more good-sized capacitors filters the supply to the drivers and preamplifiers through a series voltage regulator.

Short-circuit protection is provided by a 4-ampere slow-blow fuse in the power transformer primary. Each output stage has a thermal relay connected in the dc supply line to the top transistor. If the output current goes above a preset limit, the relay opens, breaking the power supply circuit to that stage.

Fig. 8-5 shows the reverberation board circuit, which can be switched into either channel. This can be controlled by a foot-operated switch to leave the musician's hands free for playing. A vibrato circuit (a periodic variation of volume, sometimes incorrectly called tremolo, which is a periodic variation of *frequency*) is also built on its own PC board. This is shown in Fig. 8-6. This, too, can be controlled by a foot-switch.

The preamplifiers and special-effects units are built into a small separate cabinet mounted on the top of the large speaker cabinet. The power amplifiers and main power supply are mounted on a separate chassis, which actually forms the bottom of the speaker cabinet. Fig. 8-7 shows this and the speaker cabinet. It stands almost 5 feet high, about 30 inches in width, and is set on four good-sized caster wheels.

The two big speakers are mounted on a baffle board inside the cabinet, as shown, with a protective grille and framework in front of the cones. Mounting holes for tweeter horns are provided in some versions. If they aren't used, these holes are sealed, to provide a better air-cushion for the speakers. Small "ports" are seen between the speaker holes, making this enclosure more of a "bass-reflex" design. This gives a controlled feedback from the back of the speaker cones and lowers the fundamental bass-response of the enclosure.

Because of the power output used, this whole cabinet is built out of very heavy wood, usually ⅝ to ¾ inch thick plywood. All joints are glued and held by heavy screws to avoid "drumming," or flexing of the sides at full power. The back is also made of thick plywood and held on by a great many screws for rigidity. A heavy cross-brace across the middle of the cabinet adds more stiffness to make the cabinet more solid. Handholds are cut into the slides to make it easier to carry. (It is *not* light!)

TEST AND ADJUSTMENT PROCEDURES FOR MAGNATONE PS-300 AMPLIFIER

Most factory service data give a special procedure for checking these high-power types to make certain that they will deliver the full rated output. Here is the test method for the amplifier of Fig. 8-4. Refer to the figure for location of controls and parts in the

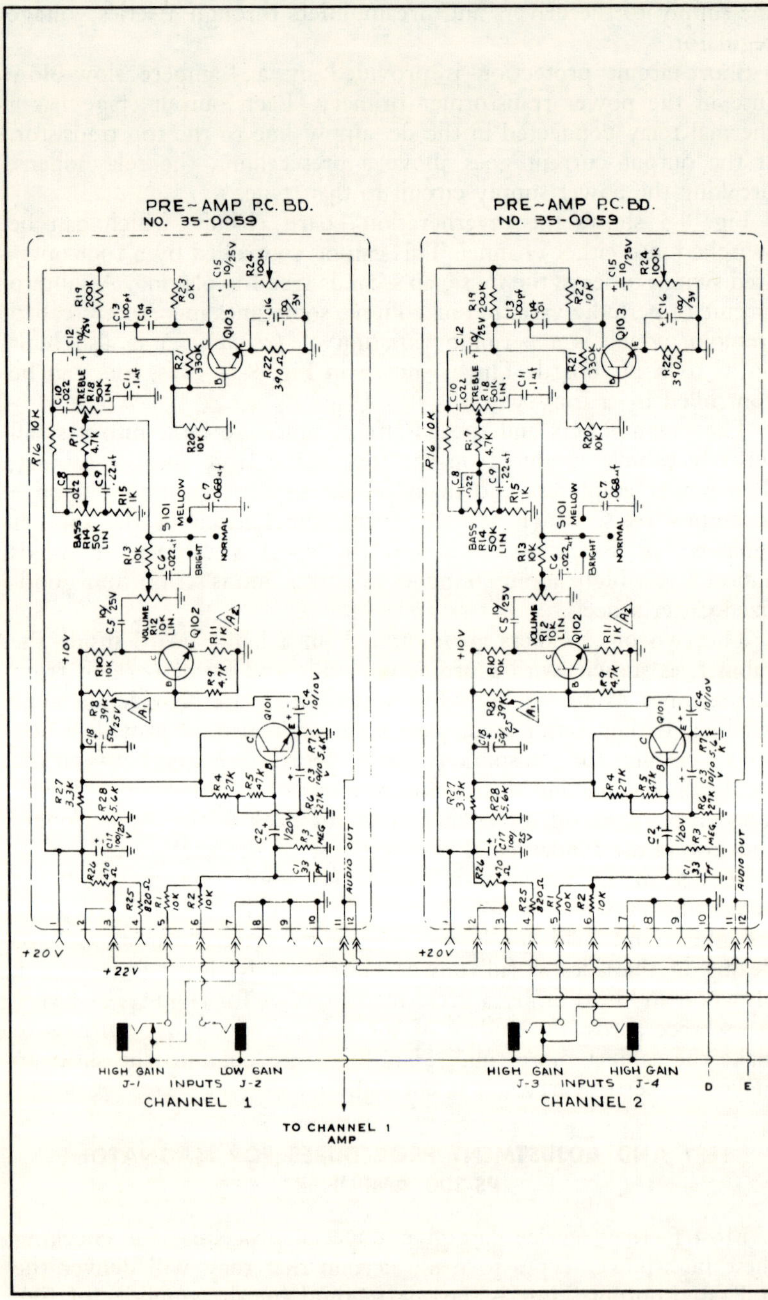

Fig. 8-4. Magnatone

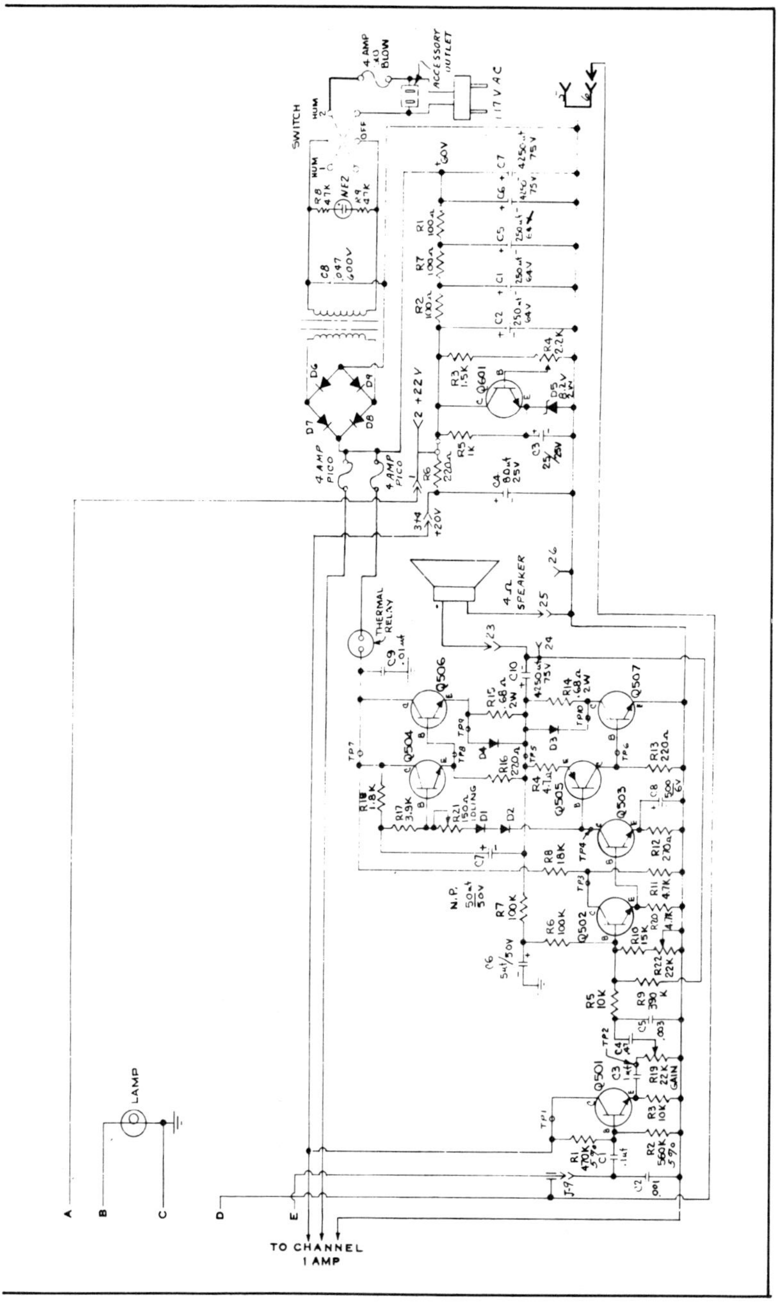

Courtesy Estey Musical Instrument Corp.

PS-300 bass amplifier.

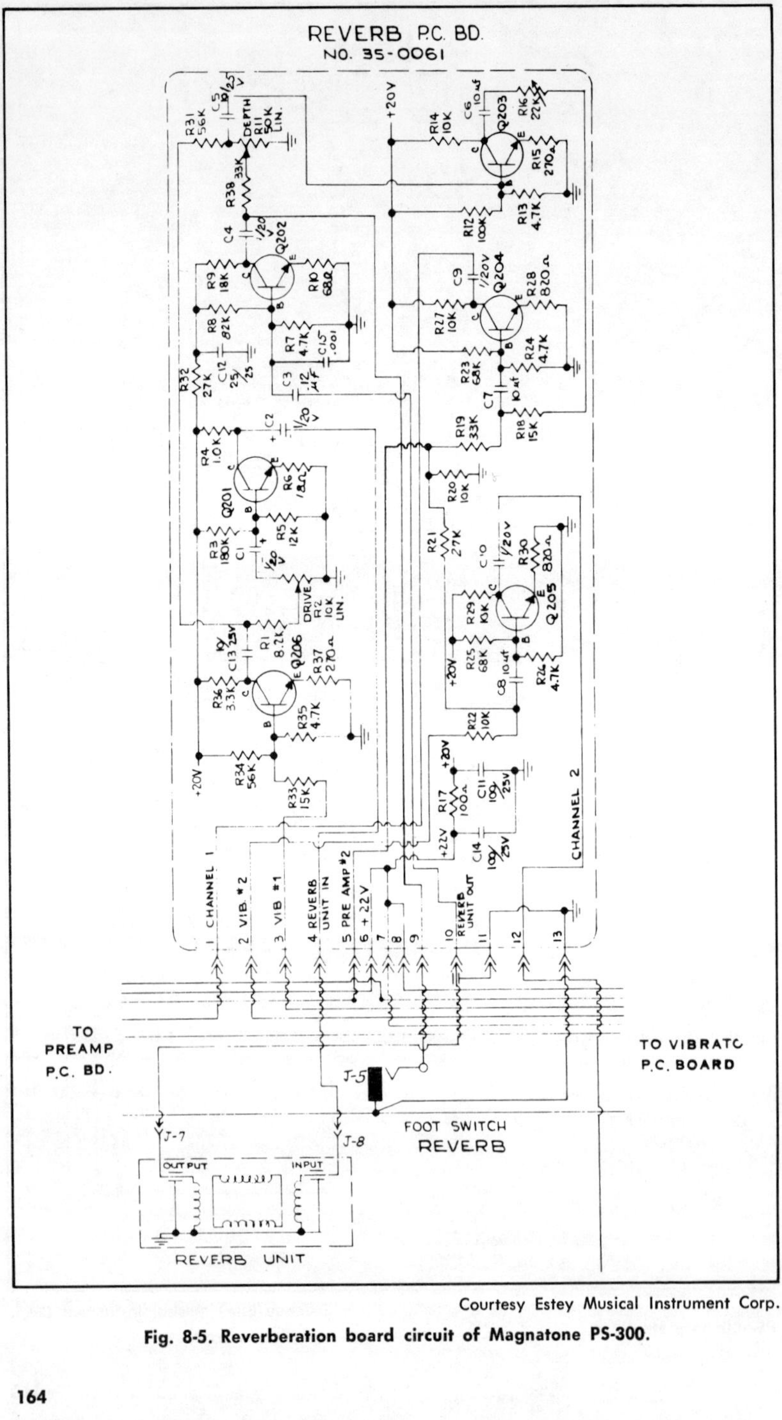

Courtesy Estey Musical Instrument Corp.

Fig. 8-5. Reverberation board circuit of Magnatone PS-300.

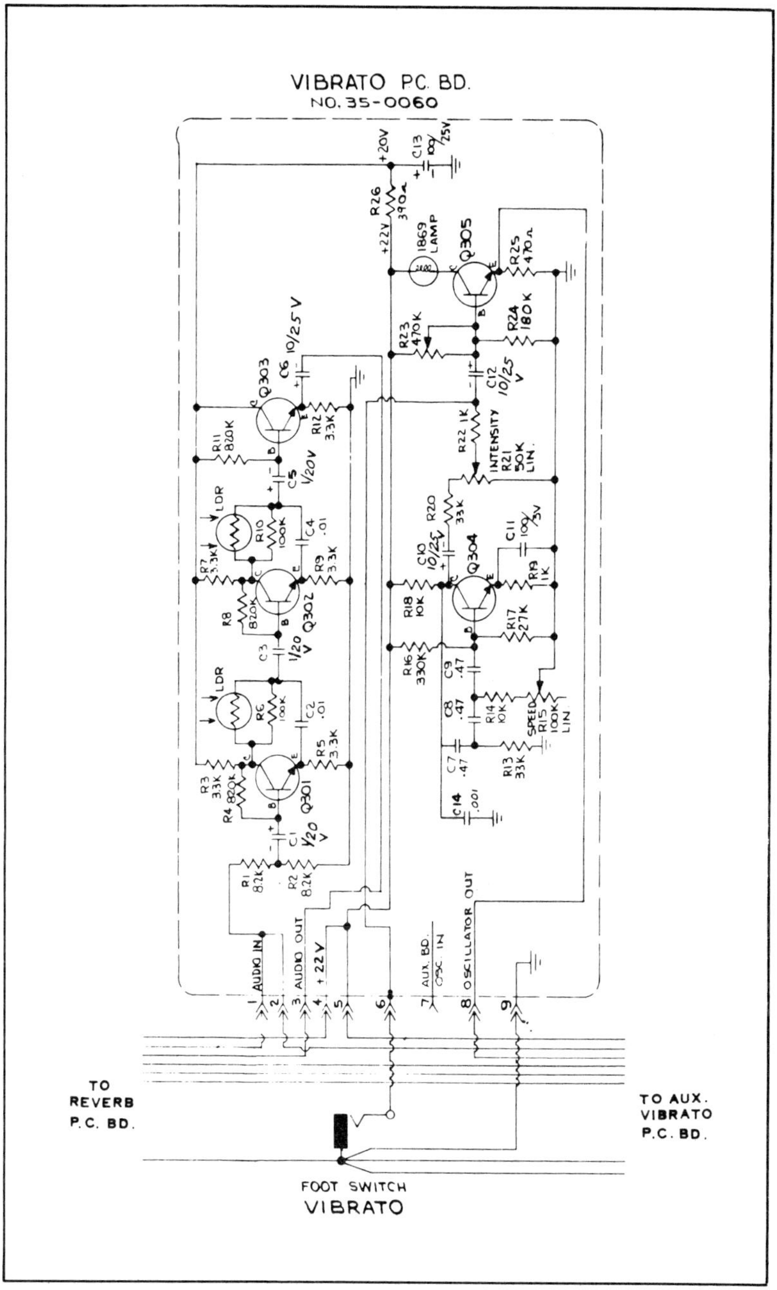

Courtesy Estey Musical Instrument Corp.

Fig. 8-6. Vibrato board circuit of Magnatone PS-300.

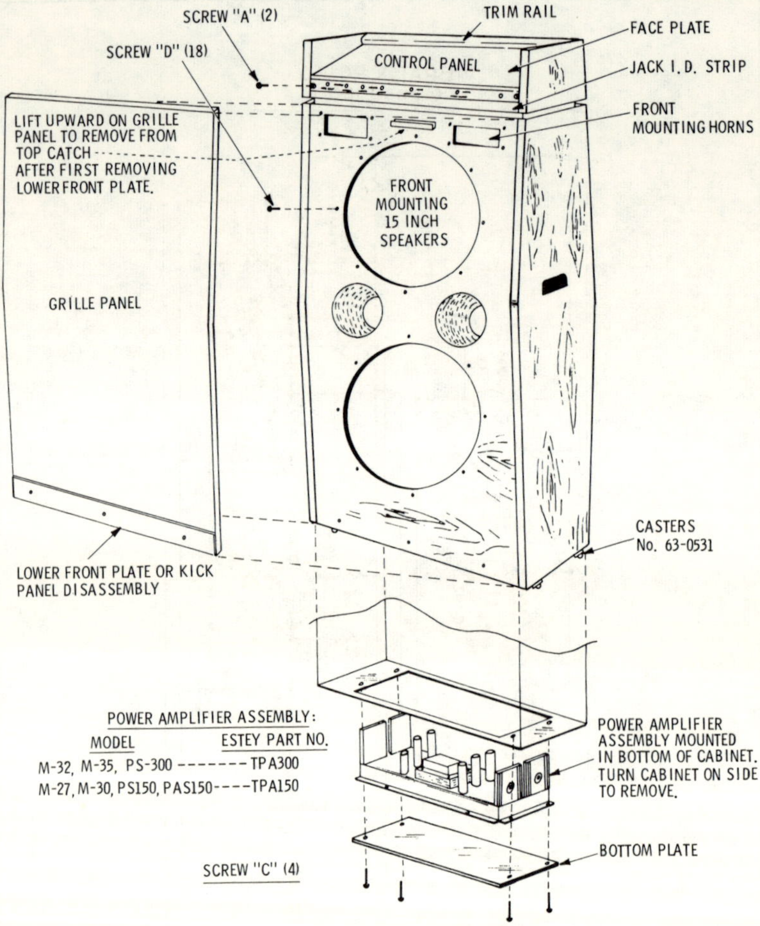

Fig. 8-7. Cabinet and speakers, Magnatone PS-300.

circuit. The procedure is the same for both channels, so we'll use the right one. (Part numbers in each channel are identical; for instance, the top transistor in each output pair is Q506, and so on.) Here it is:

1. Check all dc voltages, with everything connected. You should get +60 volts on the collector of the top transistor, and the voltage regulator output should be +22 volts.
2. Turn the power off. Remove the speaker leads, which are fitted with push-on connectors.
3. Disconnect the input cables, which are small coaxial cables with "RCA" phono plugs.

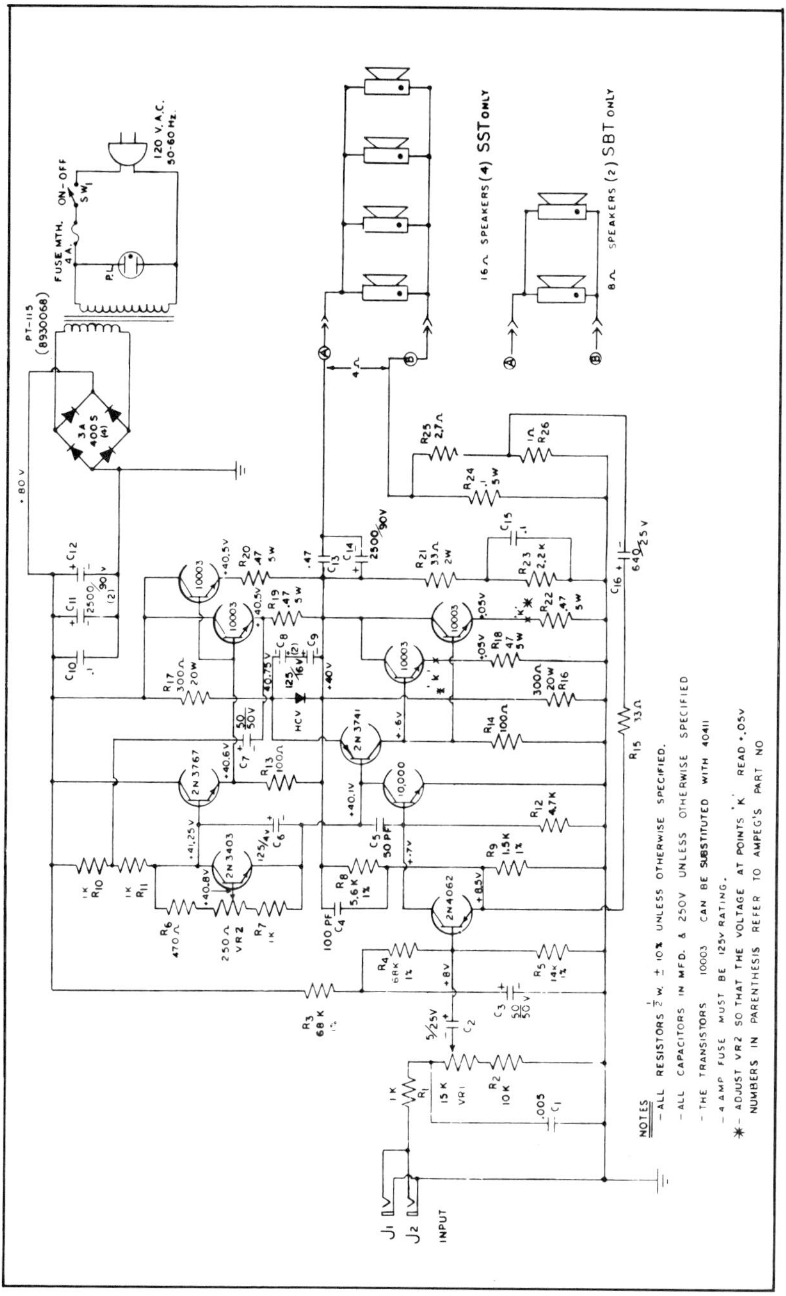

Courtesy Ampeg

Fig. 8-8. Ampeg SST, SBT guitar amplifier.

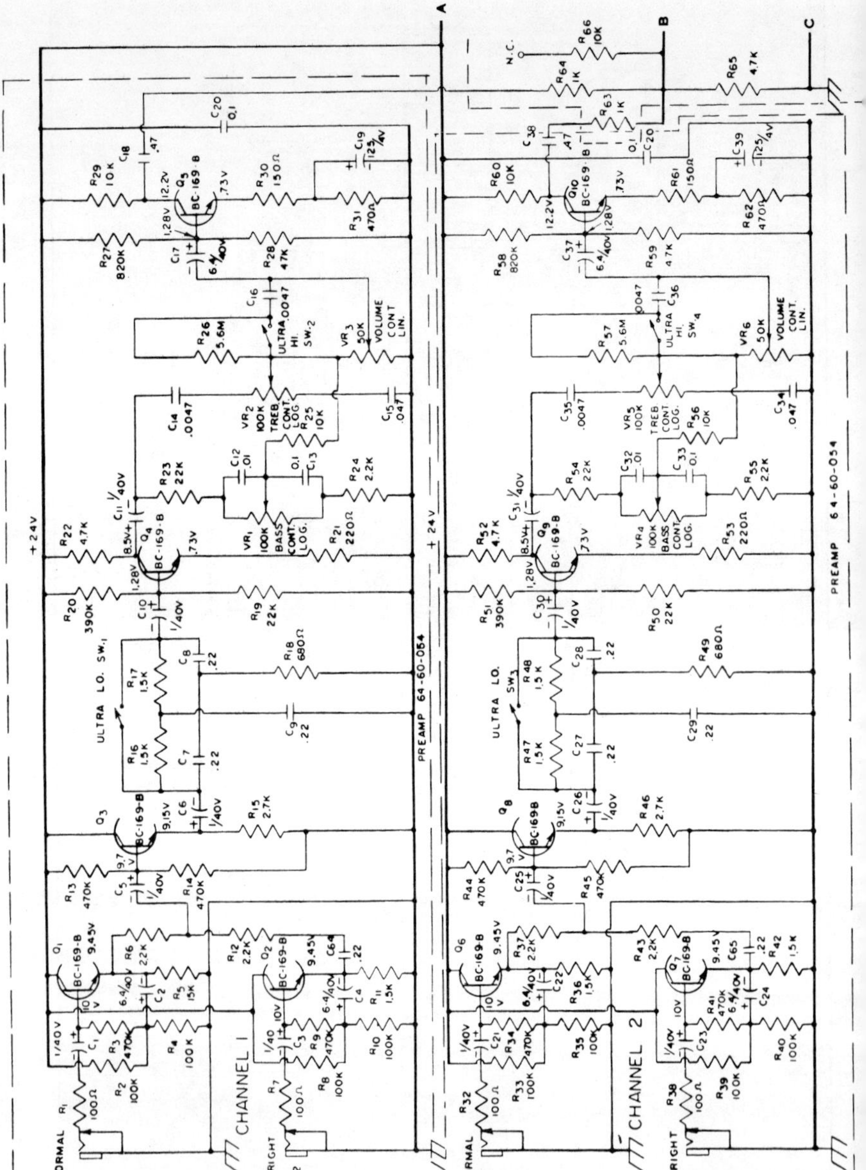
CHANNEL 1
CHANNEL 2
PREAMP 64-60-054
+24V
NORMAL
BRIGHT
ULTRA LO. SW1
VOLUME CONT LIN.
N.C.
A
B
C

Fig. 8-9. Ampeg

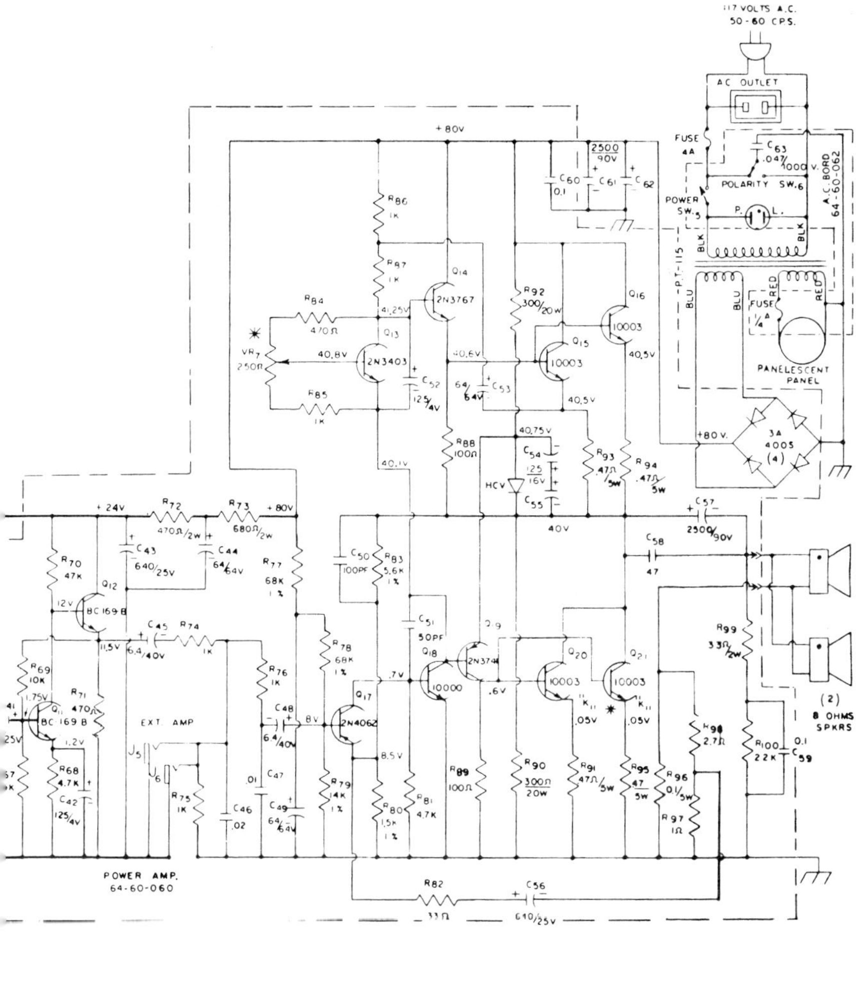

NOTES 1- ALL RESISTORS ½W & 10% UNLESS OTHERWISE SPECIFIED.

2- ALL CAPACITORS IN MFD. " " "

3- D.C. VOLTAGE READINGS WITH NO SIGNAL INSERTED, TAKEN WITH VTVM.

4 THE TRANSISTOR BC-169-B CAN BE SUBSTITUTED WITH THE 2N3391A

" " 10000 " " " " " 40408

" " 10003 " " " " " 40411

✱ 5 ADJUST VR_7 SO THAT THE VOLTAGE AT POINTS "K" READ .05 V.

Courtesy Ampeg

BT-140 amplifier.

4. Connect a 4-ohm, 100-watt resistor across the speaker leads of the channel being checked.
5. Connect the scope and an ac vtvm across this resistor. (This being an OTL pair, no harm will be done if the leads come loose, but don't let them short to each other!)
6. Feed in a 500-Hz 800-millivolt rms test signal to the input jack of this channel. (Note the lower test frequency, since this is a Bass amplifier.) Turn the power on.
7. Adjust the Gain control (R22) until the signal on the scope screen just shows clipping—flattening of tops and bottoms of the waveform. Adjust the scope vertical gain to keep the signal on the screen; don't bother to calibrate the scope to read voltage!
8. If the signal shows unequal clipping; that is, if the tops are clipped and the bottoms aren't, adjust the Centering Control (R19), which is in the base circuit of the Pre-Driver transistor Q502, until you get equal clipping of tops and bottoms.
9. Back off the Gain control (R22) until you get maximum gain without clipping.
10. Now, read the ac voltage on the vtvm. This should be a minimum of 16 volts. For a maximum power output, this would be 20 volts, indicating a power output of 100 watts undistorted.

The same tests are then used on the other channel.

Idling-Current Testing

If it's not possible to get enough undistorted power output, check the idling current of the power transistors. If this isn't set to the right point, it won't be possible to drive the amplifier to full output without running into severe crossover distortion. This adjustment checks and sets the crossover-point bias; it should not be zero, but a certain low voltage to put the output transistors at exactly the right point on the curve for correct operation. Here's the method:

1. Disconnect the speaker leads from the channel under test. Leave them open.
2. Pull the input cables from the preamplifiers.
3. Disconnect the collector lead from the top transistor Q506. This is another push-on lead.
4. Connect a 0-100 dc milliammeter between the collector lead and the case of the transistor. Connect the positive lead of the milliammeter to the +60 volt lead, negative lead to the heat sink of Q506. (CAUTION: Do NOT let the collector lead touch the chassis or any other components. Even with power turned

off, there may be ample charge left in those tremendous filter capacitors to blow Q504 or damage other transistors.)

5. Turn the power on; no signal input, no load on the output.
6. Adjust the "Idling Control" (R21) in the base circuit of Q504, for a meter reading of 50 milliamperes. Check the ac line voltage, too. This setting should be made with an ac input of exactly 117 volts. If your power line fluctuates slightly during the tests, it will cause this reading to change with it.
7. Turn amplifier off, disconnect test equipment, and reconnect speaker and input cables. Repeat the test on the other channel to make sure that they are matched.

Speaker Phasing

If you disconnect one speaker in these amplifiers, or replace one, be sure that the two are in-phase when you reconnect them. In original speakers, or exact replacements, you'll find a phasing mark on one of the terminals. This may be a dot of paint, or in this case with the Magnatone amplifier, a + sign. Be sure that the phasing-mark is connected to the *same* terminal on each amplifier. One speaker terminal goes to the midpoint of the amplifier output circuit (the hot lead), and the other to ground or common. It makes *no* difference which one goes to what, but *both* must go to the same terminal, or the speakers will not be in-phase. Since there is nothing but pure ac in the speaker circuits, there's no actual "polarity" as in a dc circuit. However, both speaker cones must move in or out *together.*

If they are out of phase, one moving in and the other out, you'll get a very peculiar distortion. This will be most noticeable at a given distance from the speakers; not so apparent up close. This is causing a partial cancellation of the sound wavefronts at some points, and a very peculiar distortion. The actual location of the cancellation-point depends entirely upon the size and shape of the room in which the amplifier is being used.

Speaker Rattles

With the high power developed in these speakers, everything must be very solid. The sound vibrations can loosen bolts or screws, and cause rattles and buzzes to show up. If the user complains of this kind of sound while playing, check the amplifier under at least partial output. The fastest way to find these is by feeding in the output of a variable-frequency audio signal generator, and running it over the whole range of audible frequencies.

Practically all of these will be some sort of "harmonic vibration"; the loose part will buzz or rattle at one given frequency and at all harmonics of this frequency. The signal generator will give you a "single-signal" test, and the rattle will be easy to find. Turn the generator dial from a low frequency up to a higher one, very slowly, and listen for the rattle or buzz. Most of them will show up at a low frequency.

Once you find it, leave the signal-generator on this frequency and investigate the cabinet. Look for loose bolts and screws, broken glue-joints, or small pieces of plywood which have become loosened, perhaps due to moisture, accidental dropping of the case, etc. You'd be surprised what a loud buzz a single splinter can make! Accidental damage to a speaker cone can also cause buzzing or rattling. Take the grille assembly off, and check the cones, especially around the outer rim. They have been known to break and tear loose at this point, which is the point of maximum flexure.

If a speaker cone is torn "axially" (a small straight tear in the cone, pointing at the voice coil), you can sometimes repair this with a *thin* coating of epoxy cement on both sides of the cone. Use as little cement as possible, and let it dry at least overnight. Radial tears, that is, cuts or breaks "around" the cone in line with the rim, are more difficult, but if they're caught in time, they can sometimes be satisfactorily repaired with epoxy. *Do not* attempt to use any kind of patch or tape on these cones. It won't last, and will soon come loose and make a worse buzz than the original tear. If the cone is badly damaged, send the speaker back to the maker for re-coning.

Many noises originate in loose parts of the speaker cabinet: cross braces, sides, panels, broken plies of plywood, etc. In one notable instance, a loose caster-wheel *under* the cabinet was resonating at a low frequency and making a very peculiar sort of "ringing" noise, like a small bell. Don't overlook *anything*. Loose mounting bolts on the speaker will also be a common cause of vibration. If anything has come loose inside the cabinet, it can be re-glued. Use plenty of glue, and if possible, add a few extra screws to hold the part firmly in place. Incidentally, if the noise is due to the original construction of the cabinet out of too-thin plywood, extra panels can be cut and glued in place to beef this up. Cut these to a tight fit, use plenty of glue, and clamp the reinforcement in place until it is good and dry, at least overnight. Extra cross braces, etc., can be added to stiffen the cabinet if needed.

THE AMPEG HIGH-POWER AMPLIFIERS

Fig. 8-8 shows the circuit of a professional-type high-powered guitar amplifier, the "Ampeg" models SST and SBT. This is a single channel amplifier, capable of delivering around 125 watts of undistorted power. Note the use of paralleled power transistors, Type

10003 (or RCA 40411) to get the very heavy currents needed to develop this much power. On the SST version, four 16-ohm speakers in parallel are used. On the SBT, two 8-ohm speakers, also in parallel, are used. The amplifier works into a 4-ohm load impedance in both cases.

Fig. 8-9 shows the full schematic of a Model BT 140, with all of the preamplifiers and special-effects circuits. The BT-140 is a two channel type, with a total power output of 250 watts, once again into 4-ohm loads. Paralleled 1003s are used in the OTL circuit, driven, direct coupled by two driver stages, in a quasi-complementary-symmetry circuit. (Only one power output channel is shown, since the other is identical.)

TROUBLESHOOTING THE GIANTS

These big amplifiers can be tested and repaired with the same methods you use on the smaller ones. The only difference lies in the greater load currents and higher voltages in some of them. The first step, of course, is to check for normal dc power supply voltages. Second, check the signals through the whole circuit. Third, check parts and transistors in any circuit that won't pass a signal or that distorts it.

The resting-current test is very useful. This will immediately spot any leakage in the output transistors. You'll find that the idling-current control won't bring the collector current at no-signal down to the normal level. Check for matched idling currents on both channels. This trouble can also be in one of the driver transistors, in direct-coupled circuits, which is upsetting the bias on the whole stage.

As usual in direct-coupled circuitry, once you do know there is trouble in it, take it apart. The only reliable way of making tests in these circuits is to break the circuit somewhere near the middle, by taking out a transistor, etc., so that you can get valid tests on the others. Transistors in these must be tested out-of-circuit.

If you find troubles such as blowing the fuse, check the power supply first with the output transistors disconnected. If this is working, then check the power transistors, driver transistors, and all parts in this stage. Even things such as an open bias potentiometer or bias diode can upset the circuit enough to make it blow fuses.

CHAPTER 9

Replacing Small-Signal Transistors

A lot of the service jobs you do on transistor amplifiers will involve locating defective transistors, finding a suitable substitute, and replacing them. Physically, this isn't too hard, though it can be in the more crowded chassis. We'll give you some hints on making that part of it easier, too.

The most difficult part of the job will be finding a good replacement for the bad transistor. In audio work, this isn't as bad as in other applications, such as rf and i-f, sync clippers in TV, etc. We do not have to bother with the high-frequency cutoff of our audio transistors. This will always be well above any frequency we have to handle in this type of work. Twenty-five kHz is plenty; most will go up to at least 50 kHz, and many to 1.0 MHz. In a typical list of replacement transistors, the lowest cutoff frequency of any is 210 kHz, and this is a very big transistor. Smaller ones go higher.

MATCHING CHARACTERISTICS

All transistors are "triodes," (with the exception of a few special types such as FETs, etc.) All of those we will find will be triodes. So, this will be a lot easier than finding a replacement tube. All we have to do is find a replacement that will match the important characteristics of the original as closely as possible. Contrary to many statements you've seen in the past, this is not hard to do.

Transistors have an awful lot of "parameters," as you've found if you've read any of the "engineering" type of transistor books. Besides the voltage and current parameters, there are the leakage parameters, temperature, 'h' parameters, 'y' parameters, and undoubtedly several others that I haven't found yet. These things are an "engineer's delight"; they can cover page after page with mathematical expressions on how to find the parameters. This is fine, if you're *building* transistors, or if you're designing equipment, but we're not. We have, in all cases, a circuit that is already built, and which did work at one time! So, all we have to do is check up on a *very few* of the most important characteristics, look these up in a list, and there we are. It's not any harder than buying a new tire for a car, if you know the size of the worn-out one.

We need to know five things about a replacement transistor. Here they are:

1. Type. Npn or pnp?
2. Material. Germanium or silicon?
3. Maximum collector-emitter breakdown voltage; (BV_{ceo}).
4. The current gain or beta.
5. The type of case. (Although this is not nearly as important as the first four; we can always "convert.")

"Getting in the Ball Park": Parameter Spreads

Here are a few facts about transistors that aren't too widely known or publicized except among transistor makers and designers of solid-state equipment. We can use these to very good advantage. First, transistors are not as critical as they're supposed to be. Unlike vacuum tubes, where all tubes of the same type-number have characteristics that match within a very small tolerance, you can have a very wide *variation* in operating characteristics of transistors *of the same type number*. This is the "nature of the brutes."

For example, you'll find three columns under the heading "beta" (or h_{FE}) in transistor data sheets: "Minimum, Maximum, and Average." If the transistor has an average beta of 100, and you bought 100 of them, maybe 60 of them would check out at exactly 100. Twenty of the rest would read 65 to 70, and the remaining 20 would read 125 to 130.

This is called the "parameter spread," and we love it. This means that the designer must build this circuit to *work* with transistors that fall anywhere within the parameter spread, in this case, between 70 and 130. So, this gives us room to work. If you can find a replacement transistor that has characteristics anywhere "in the ball park," it will work just as well as the original.

The most important parameter, after we have chosen a transistor of the same type and material as the original, is the collector-emitter breakdown voltage. This does NOT mean that this rating must *match* the dc voltage you found in the circuit. You must exceed it, in your replacement, by a good-sized safety factor. If you have 25 volts applied to the transistor, your replacement should have a breakdown voltage rating of at least 50 volts. This safety factor gives our replacement transistors the chance to last longer. There is, as usual, a common-sense reason for this. In Class-A circuits, for example, if you have an applied voltage of 25 volts, remember that you'll have a signal-voltage *swing* of perhaps twice this. Unless the dc voltage supply is very well regulated, you could. So, don't take chances with too low a value; use a replacement with the highest possible breakdown voltage.

Transistor Numbering

Right at this moment, transistor type-numbering is one of the most confused areas in the world, outside of politics. Although the EIA made a mighty effort to bring order out of chaos, they may as well give up, it seems. Consequently, when you open up any piece of transistorized equipment, the chances of seeing any transistor in there with a number you recognize is very small—almost nonexistent, in most cases.

There are three numbering systems. One, the EIA system, is used on American-made transistors, and only on transistors which have been "type-accepted" by the Semiconductor Registration Division. These will have numbers beginning with "2N—" etc, for transistors, "1N—" for diodes, and so on. The Japanese Standard, equivalent to our EIA, uses type-numbers which begin with a number and two letters, such as "2SA—," "2SB—" and so on. (They're up to "2SD—" as this is written, and they may go on.) These are as near to a recognized and accepted standard as you'll get in imports, so make use of it and enjoy it.

"In-House" Numbering

The most confusing of all is "in-house" numbering of transistors. These are transistors which have been made to specifications of the manufacturer of the equipment, and assigned arbitrary, meaningless numbers, according to his own code. Everyone seems to use a different coding, too. There are so many different in-house systems that no one could remember them, so don't bother. Once in a while, you'll find one which makes at least a little sense, such as Magnavox's use of the letters "N" and "P" to denote npn and pnp transistors. Thus, a "63N1" is an npn, and so on. After a little practice, you will be able to recognize a few of the major manufacturer's transistors by their characteristics numbers. Don't sweat it, though. There are easier ways.

THE GENERAL REPLACEMENT LISTS AND CATALOGUES

Up to this point, the picture has been pretty gloomy. Cheer up; things aren't as bad as they've seemed to this point. There is help for us, and it's a good thing. Without it, we would be literally "help-less." I'm referring to the American transistor manufacturers who publish "General Replacement Guides." These are literally worth their weight in germanium.

These listings are published in the Howard W. Sams *Transistor Substitution Handbook* and by U.S. semiconductor makers who have a special line of "General Replacement" transistors. The Guides are published at intervals, as the list of new transistors gets larger. In one of the latest editions, some *28,000* transistor "types" are listed!

Here is what might sound unbelievable, but is demonstrably true. Everyone of these thousands of transistors can be replaced by one of a group of not more than *50* replacement types. The transistor manufacturers have lines of general-replacements that will match the characteristics of practically any transistor you can find, including a lot that won't be listed in the books. The originals will have a given set of characteristics, and you can always find a replacement that will be in the ball park, and work satisfactorily.

Some time ago, I conducted a long series of tests to verify this. I actually made replacements of transistors in amplifiers, radios, and tv sets, both B/W and color, sometimes using the listings in the Guides, sometimes using the method outlined in this book. (Even where the originals were listed, this method came out to show the type of transistor listed.) In every case, the replacement transistors worked as well as the originals. The performance tests, of waveforms, dc voltages, currents, etc., were taken "before and after," and showed that these replacements are satisfactory in every way.

So, when you find a bad transistor in an amplifier, look up the number in the Substitution Guide, if possible, and use the recommended type from the list. If you can't find the number in the Guide, make up the list of characteristics as given before, then go to a Characteristics Chart such as the Howard W. Sams *Transistor Specifications Manual*, select one that is as close as possible, and there you are. Now let's see how this can be done.

FINDING THE CHARACTERISTICS OF UNKNOWN TRANSISTORS

In this section, we're dealing only with the small-signal transistors, preamplifiers, drivers, etc. It will be comparatively simple to make replacements of these. Power transistors require more care, and they'll be dealt with in the following chapter.

The most helpful thing is having the service data, the schematic diagram with voltages and currents given on it. Sams PHOTOFACT schematics do include this data, taken directly from actual samples of the amplifiers. For some reason, the schematics of some manufacturers do not give load currents, but do give the dc voltages. It can be done without this data, but it's harder.

Now, let's replace a transistor, using this method. You have found a certain stage not working. It refuses to pass a signal, and its dc voltages are far off normal. Removing the transistor, you use your transistor-tester to confirm the diagnosis. It's bad. So, we need a replacement.

The easy way: Look it up in the parts list of the Sams PHOTOFACTS. Here, the correct replacement in several different manufacturer's lines will be given. That's one way.

The next harder way: We have the schematic and voltages, but no identification as to numbering, etc. But, we do have the type, pnp or npn. Make up a scratch paper list. To find out which material is used, look at the bias. If this is about 0.2 volt between base and emitter, the original is a germanium type. If it's about 0.6 volt, it's silicon. That's two facts.

Next, find out the maximum dc voltage that will be applied to this transistor. Look on the schematic. Check the dc voltage of the power supply. Last resort, (transistor out) turn the power on, and read the highest dc voltage at any of the empty holes. That's three facts.

Now, let's say that this is an npn silicon transistor, and it had 16 volts on its collector. Say this was used in an early stage as a preamplifier. If so, it will have a medium beta, perhaps 60 to 90 or so. There are all the facts you need. Go to the Transistor Replacement Guide, and in the charts look for an audio transistor with these characteristics. Fig. 9-1 shows one of these charts, illustrating what a wide range of types we have to choose from. Our transistor, with the characteristics given, could be replaced by a Motorola HEP-53, RCA SK-3038, Sylvania ECG-123, or GE-10.

The least important of all is the case, and the basing. There is no standard for transistor bases, not yet. You'll find E-B-C, E-C-B, B-E-C, and so on. However, since all of them (in this class) have three wire leads, you can bend them around in any way necessary to get them through the holes in the board. If this requires too close a twist, slip pieces of very thin spaghetti over the wires to keep them from shorting to each other or to other things nearby. Strangely enough, reversing a couple of leads doesn't seem to damage modern transistors, but it certainly doesn't help the gain of the amplifier. If you have made a replacement, and suddenly the amplifier doesn't work at all, go back and check your own work. In one case, I had to try *three times* before I got all of the little wires in the right holes.

SILICON TRANSISTORS

MINIMUM RATINGS AND CHARACTERISTICS ($T_A = 25^oC$)

LOW POWER

HEP Type #	Polarity	Collector-Base Voltage BV_{CBO}	Collector-Emitter Voltage BV_{CEO}	Emitter-Base Voltage BV_{EBO}	Collector Current I_C (mAdc)	Total Device Dissipation (mW)	Operating Junction Temperature (°C max)	Typical Cut-Off Frequency f_t (MHz)	Typical Current Gain h_{FE} (β)	Case
50	NPN	25	15	4.0	300	400	175	250	85	TO-18
51	PNP	30	25	4.0	600	600	175	150	80	TO-5
52	PNP	30	30	4.0	200	400	175	200	95	TO-18
53	NPN	35	30	4.0	600	600	175	200	85	TO-5
54	NPN	30	20	5.0	200	310	135	30	350	TO-92
55	NPN	30	25	5.0	200	310	135	200	350	TO-92
56	NPN	30	20	2.5	100	310	135	750	70	TO-92
57	PNP	25	25	5.0	200	310	135	200	350	TO-92
709	NPN	30	15	3.0	50	300	175	600	120	TO-72
715	PNP	50	40	4.0	200	310	135	200	120	TO-92
716	PNP	40	40	5.0	600	310	135	300	110	TO-92
717	PNP	25	25	4.0	100	310	135	120	350	TO-92
718	NPN	30	20	4.0	100	310	135	600	80	TO-92
719	NPN	30	30	4.0	100	310	135	700	130	TO-92
720	NPN	30	20	3.0	100	310	135	800	40	TO-92
721	NPN	30	30	5.0	500	310	135	260	350	TO-92
722	NPN	25	25	4.0	100	310	135	200	60	TO-92
723	NPN	25	25	4.0	100	310	135	200	90	TO-92
724	NPN	25	25	4.0	100	310	135	200	160	TO-92
725	NPN	25	25	4.0	100	310	135	200	250	TO-92
726	NPN	25	25	4.0	100	310	135	200	350	TO-92
727	NPN	25	20	4.0	100	310	135	200	50	TO-92
728	NPN	45	40	4.0	100	310	135	200	180	TO-92
729	NPN	45	45	4.0	100	310	135	200	100	TO-92
730	NPN	30	25	4.5	50	310	135	175	600	TO-92
731	NPN	20	20	4.0	25	310	135	30	130	TO-92
732	NPN	20	20	4.0	25	310	135	200	45	TO-92
733	NPN	30	20	4.0	100	310	135	100	95	TO-92
734	NPN	40	20	4.0	25	310	135	200	75	TO-92
735	NPN	45	40	6.0	600	310	135	150	300	TO-92
736	NPN	55	50	6.0	600	310	135	150	140	TO-92
737	NPN	30	25	6.0	100	310	135	100	600	TO-92
738	NPN	45	40	7.0	100	310	135	100	250	TO-92
739*	PNP	40	35	25	150	310	135	4.0	250	TO-92

* Silicon replacement for 2N404.

RF POWER

HEP Type #	Polarity	BV_{CBO}	BV_{CEO}	BV_{EBO}	I_C (mAdc)	Dissipation	Temperature (°C max)	f_t (MHz)	h_{FE}	Case
75①	NPN	40	20	3.5	400	3 W	175	250	15	TO-39
76①	PNP	40	20	3.5	400	3 W	175	250	15	TO-39

① These devices may be used as complementary pairs.

MEDIUM POWER

HEP Type #	Polarity	BV_{CBO}	BV_{CEO}	BV_{EBO}	I_C	Dissipation	Temperature (°C max)	f_t (MHz)	h_{FE}	Case
242①	PNP	60	40	7.0	3 A	6 W	200	8.0	60	TO-5
243①	NPN	60	40	6.0	3 A	6 W	200	8.0	60	TO-5
245②	NPN	60	40	5.0	3 A	30 W	150	8.0	60	77
246②	PNP	60	40	5.0	3 A	30 W	150	8.0	60	77
700	PNP	40	40	5.0	5 A	40 W	150	8.0	70	77
701	NPN	40	40	5.0	3 A	40 W	150	8.0	100	77
702	PNP	80	80	5.0	3 A	25W	150	8.0	80	TO-66
703	NPN	70	60	5.0	5 A	25W	150	6.0	85	TO-66
705	PNP	40	40	5.0	5 A	87.5 W	200	4.0	250	TO-3
708	PNP	60	60	5.0	600	1.8 W	175	275	85	TO-18
710	PNP	95	80	4.5	100	3 W	200	200	25	TO-5
712	NPN	180	200	5.0	50	1 W	175	150	35	TO-5
713	NPN	100	100	5.0	200	1 W	175	150	35	TO-5
[illegible]	[illegible]	[illegible]	150	5.0	[illegible]	[illegible]	[illegible]	[illegible]	120	[illegible]

Courtesy Motorola Semiconductor Products Inc.

Fig. 9-1. Motorola HEP Transistor Specifications.

In another, after replacing a transistor, I got a terriffic distortion; the waveform had to be seen to be believed. A recheck showed that I had been careless in reading the type number of my replacement transistor, and I had a pnp in place of an npn. All you can do is grin sheepishly, take the wrong one out and put the right one in, double-checking its type this time.

EXACT-DUPLICATE REPLACEMENTS

Of course, if you do have exact-duplicate replacement transistors from the manufacturer (if it is an American company) use them. If you need something called a "Device 224" and you've got some on the shelf, use them. Saves all of the work of going through the replacement lists. This is especially useful in the case of high-power replacement output transistors, as you'll see in the next chapter. You'll be able to eliminate all but the most necessary performance tests.

As we said earlier, watch out for the single-case Darlington-pair transistors. So far, these will be found mainly in preamp and driver stages, where their high gain is very useful. One manufacturer is building "power Darlingtons," with an output of about 2 watts. So, this can be the whole amplifier. At present, at least, these are built in what look like TO-3 cases from the top. On the underside, you'll see not two, but eight terminals, and one quick look will be all you need to identify them.

If a Darlington preamp-driver is out, you can make up a good replacement out of two identical transistors of the same type, pnp or npn (generally silicon, but the circuit will work with either type, of course). You can twist the leads together in the *Darling*ton circuit and solder them in the correct holes. (Be sure to insulate the emitter-1/base-2 lead, since it does *not* "come out" to the circuit.) If the original was used as a driver, direct coupled to the base of the output pair, be certain that this base voltage on the output is correct. If this voltage is too high, try using transistors with a lower beta for the replacement; if it's low, use higher-beta types, until you get as close to the original bias as you can. For a definite check, be sure that the output waveform isn't clipped.

DRIVER TRANSISTOR REPLACEMENT

Preamplifier transistors ordinarily use very low power types, so you won't have to worry about that. However, if the defective transistor is a driver, especially in amplifiers from about 5 watts on up, check the power rating. Maximum power in preamplifier stages will be in the order of a few milliwatts. Drivers may need transistors rated from 1 to 5 watts. Remember, most drivers work in Class A, and their average dissipation is quite a bit higher than the more efficient Class-B circuits used in the output stage.

In stereo amplifiers, you can confirm the needs by checking the actual current drawn by the same transistor used in the other channel. Be sure that your proposed replacement has a maximum current rating well above the in-circuit current drain of the "guinea-pig" in the working channel.

"FINDING THEM IN THE BOOK"

I would very definitely recommend that you get what information you can on transistor replacements and keep it up-to-date. These can be the most useful reference lists you have. In many cases, you will be able to find the type number of an unknown transistor in one of them. Cross-checking the characteristics of the replacement type given against the chart, you can tell what you need.

These listings have an amazingly high percentage of accuracy. Aside from the inevitable typographical errors, which are unavoidable but very rare in lists of this complexity, the replacements given will work. If you have doubts, though, just check the characteristics against conditions in the circuit, voltage and current ratings, for instance. This will confirm it.

CHAPTER **10**

Power-Output Transistors

Finally, we come to the "Big Boys"—the high-current and/or high-voltage types of power output transistors. Since these use up most of the power drawn by the circuits and do most of the work, they provide most of the trouble. Being under greater stress, they are more apt to fail. We need reliable tests for these, in-circuit and out, and ways of making sure that our replacements are working well inside their safe limits, so that they'll stay in there. We'll see some of these tests in this chapter.

TESTING

Basically, the tests will be exactly the same as those used before; first, the signal-check, to see if the thing will pass a signal; if it won't, then the dc voltage tests for any sign of trouble (such as missing midpoint voltages, etc.); and finally, confirmation by taking it out of the circuit and checking for shorts and leakage. At this point, we should repeat something we've said before. Germanium power transistors will have *normal* leakages of up to 100 or 150 microamperes, or even more. Silicon power transistors, on the other hand, should show practically zero leakage; in some cases, a leakage of as little as 10 microamperes has upset a stage using silicon transistors.

Since these transistors can draw high currents if they short out, they'll generally take other parts with them. When there has been a dead short in the output transistors, always remember to check such things as emitter resistors, bias diodes, and any dropping resistors in the dc voltage supply. Fig. 10-1 shows some of the things

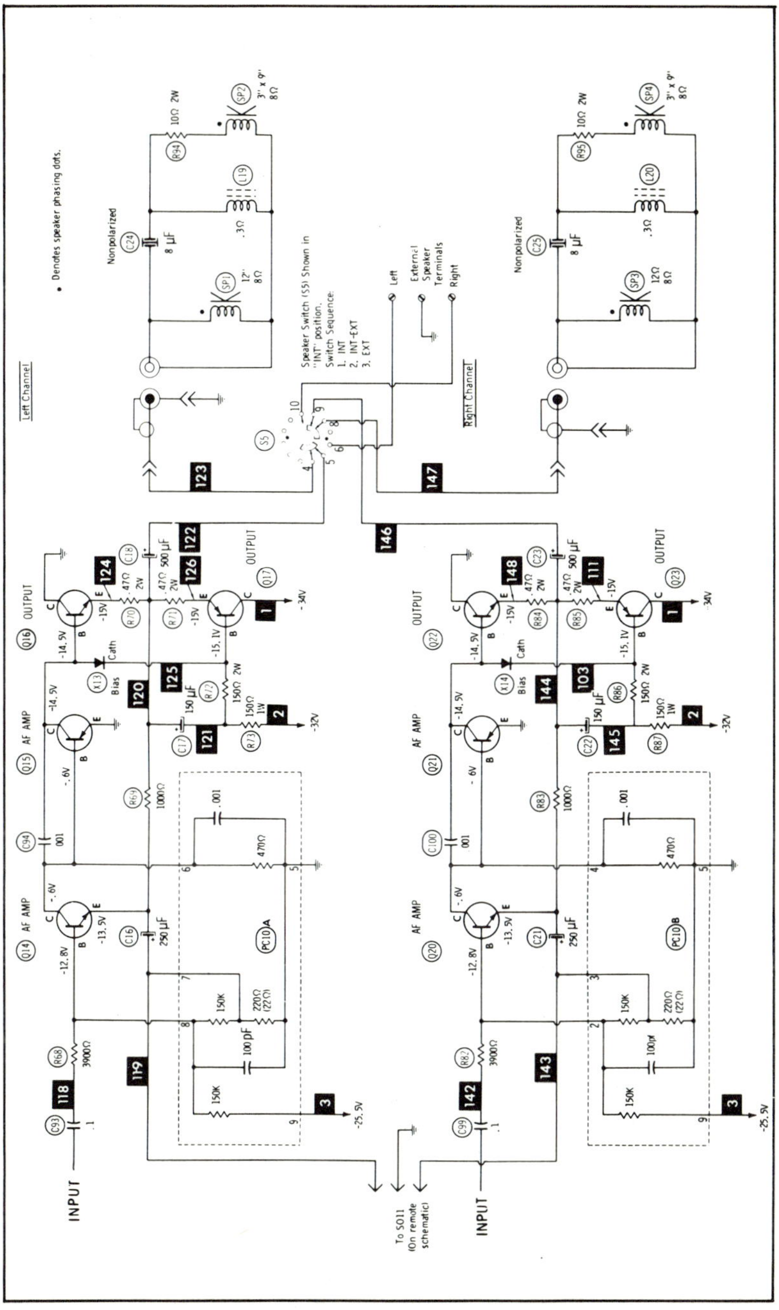

Fig. 10-1. Preliminary tests to make before replacing transistors.

that must be checked *before* making a replacement of power transistors. Silicon transistors, most especially, are apt to go to a "crowbar short" when they fail, and the high currents that flow will damage other parts. The emitter resistors often have a low wattage rating, so that they will act something like fuses and blow quickly under an overload.

VOLTAGE AND CURRENT READINGS FOR COMPARISONS

If this is a stereo amplifier, as so many are, and only one channel is bad, you can get some valuable information by taking a few key readings on the working channel. This is after the shorted transistor has been located and removed. Leave it out until these tests are finished.

Read the dc voltages on the transistors of the working channel, at no-signal and at a fairly high signal level. Break the collector circuit of the top transistor and insert the dc milliammeter. Read the resting current and make a note of it. Then, turn the volume up, with signal input, and read the full-load current. Log this. (Details on an easy way to do this will be given very soon.) Now, even without the service data, you have all of the key readings you need to select a replacement transistor type.

SELECTING A REPLACEMENT TRANSISTOR

"By the Numbers"

There is a regrettable tendency among manufacturers to use in-house numbered power transistors, identified only by a code number. In a surprising number of instances, you'll find these numbers listed in the Replacement Guides, or in the Parts Lists of PHOTOFACTS. However, let's take a "worst-case" approach and see what we can get from only the readings we got on the working channel.

In this approach, the first thing is the case. Practically all of these will be in the TO-3 cases, shown in Fig. 10-2. The older units may have the "bolt-down" TO-36 case transistors, but these seem to be quite rare lately. However, they are still available. You'll find TO-3 cases with thin flanges, and some with thick flanges. Both of these are the same. As long as the transistor will fit the socket, fine. These seem to be almost entirely bolt-ins, with sockets for the base and emitter pins.

Now all we need are the voltage and current readings. If the other channel is working, we've got them. Let's say that there is 45 volts applied, and that the maximum load current (full volume) is 800 mA. The first thing that needs to be identified is the material: germanium or silicon? We get this by checking the resting bias

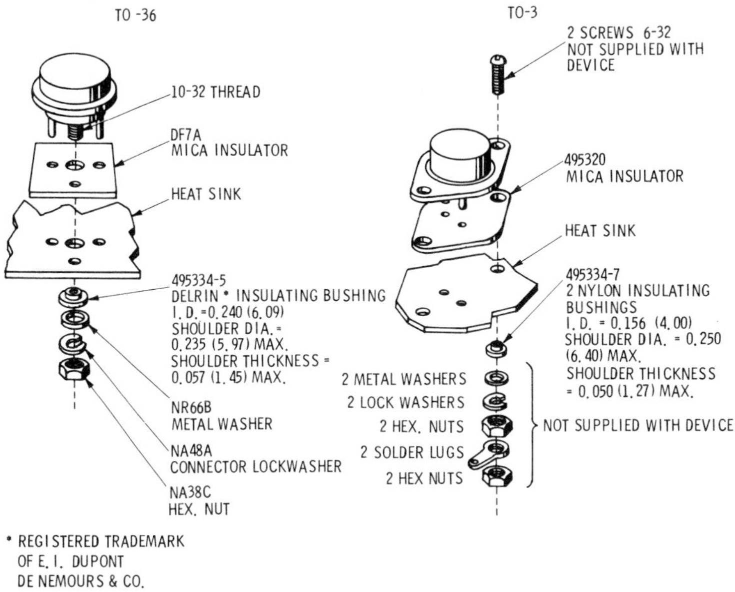

Fig. 10-2. TO-36 and TO-3 cases and mounting hardware.

between base and emitter. If this is close to 0.2 volt, germanium; close to 0.6 volt, silicon. (Within 0.1 volt either way is "close.") Then, the type: pnp or npn? By cross-checking with the working channel, we can find out whether this is a stack or a complementary-symmetry circuit.

If both of the remaining transistors have the same number, we can safely assume that this is a stack with two identical transistors. If they're different numbers (no matter what they are) they're apt to be a complementary-symmetry type. Confirm this, either with the ohmmeter or by checking them in-circuit with the transistor tester. While some of them won't check in-circuit, most of them will—at least enough to give you the *type*.

Now, we have the case (TO-3), the type (let's say silicon npn), and the dc voltage. We can start looking through the charts in the Replacement Guides. Such types as Motorola HEP-247, with a breakdown voltage of 70 volts, current max rating of 10.0 amperes, and a beta of 60, or RCA SK-3027, 90 volts, 15 amperes, and a beta of 60, would work nicely. If this circuit is a stacked pair, RCA packs matched pairs of SK-3027's under the part number of SK-3029. If it's a complementary-symmetry circuit, you'll need a pnp to match the npn. Motorola's HEP-248 is an exact duplicate of the HEP-247, but is pnp. Voltages, currents, and beta are the same.

Even if only one transistor in a pair has blown, always replace both of them. It would be a long task to find and test for a precise match with the original transistor. There are a great many matched-pair transistors available in all power and voltage ratings. Check the replacement guides, and see.

Note the nice, comfortable safety factor used in making this selection. The voltage breakdown rating is at least double the normal maximum applied voltage, and the current rating is a fine big 10 amperes at the least. Your tests on the working channel showed a maximum full-load current of less than 1 ampere, so this will be fine. If you're going to err, go to the side of higher ratings rather than lower. For example, with 45 volts applied, don't pick one with only a 50-volt collector rating. That's skating on very thin ice. Same goes for current ratings; the bigger the better.

Precautionary Tests, after Replacements

After power transistors have been replaced, be very careful when turning it on for the first time. It is always a good idea to use the Variac, as described in the chapter on Test Equipment. Turn this to zero, plug the amplifier in, with voltmeters and current meters in place, and then bring the power up very slowly. Stop at about 40 or 50 volts and see what's going on. By doing this, you can avoid the possibility of blowing your nice new pair of output transistors.

Collector-Current Tests

The rest current of the output pair is the most important "first-check." Look at your notes to see what this should be in the other channel. If this isn't available, something like 25-30 mA is a good ball park figure. Connect the dc milliammeter into the collector circuit of the top transistor. We mentioned an easy way to do this, before; TO-3 sockets usually make the collector connection through one of the mounting screws, to a spring clip under the transistor socket. The other screw just holds the transistor down. Take out the connector screw; this breaks the collector circuit. Connect the milliammeter with alligator clips to the terminal under the socket and to the collector (case) of the transistor. The alligator clip can be pushed into the empty screw-hole in most cases. As a last resort, you can hold it touching the case, with one hand.

Bring the ac input up slowly and watch the milliammeter. If there is trouble in the circuit, this current will go up to very high levels long before the Variac gets the ac input anywhere near the full line voltage. If you see the current rise to several hundred mA, STOP! Something is very definitely wrong.

One of the things which could, and does, cause this kind of problem is incorrect bias. If the output pair is directly coupled to the driver transistor, leakage or a short in the *driver* transistor can put a big forward bias on the output transistor, and may have been what blew it in the first place. You can take currents up to 100 or 200 mA for testing; read the base voltage on the output transistor. If this is an npn, the base will probably be far too much positive (forward bias); for pnp's, the forward bias will be negative, but you'll get the same result: much too great a current in the collector circuit. If the full line voltage were applied in this condition, the new transistors would probably blow in about 50 milliseconds, too fast to rescue by throwing the switch.

If the test current checks out at a low value with the full ac line voltage applied, then you can proceed with the full-load current tests. Feed in a test signal of any kind, turn the volume up slowly, and watch the collector current. In amplifiers of say 50 watts per channel output, this will probably go to about 700 to 900 mA peak. On music, you'll see the meter needle swing pretty much, but you can get an average of the current, and see about where the peak is. If you use a steady sine-wave test signal, the needle will stand still and be a lot easier to read.

For power output tests, turn the amplifier off and disconnect the speaker (or put in your ear plugs; 50 watts is a lot of audio power, especially with a sine-wave signal!). Connect a suitable high-wattage load resistor which is set to the same value as the speaker impedance, across the output. Connect the scope across this resistor, with the ac vtvm as well.

Feed in a test sine-wave signal, say 400 Hz at about 300 millivolts (good average test input), and turn up the volume until you note symmetrical clipping of the peaks of both halves of the signal. The collector milliammeter can be left in or disconnected for this test. Now, read the ac voltage developed across the load resistor at the point just before clipping is visible. Square this voltage and divide it by the value of the load resistor, and this will give you the actual undistorted power output of this amplifier. (E^2/R = Power in watts). For example, 20 volts rms across an 8-ohm load resistor shows a power output of 50 watts ($20^2 = 400/8 = 50$).

A lot of these amplifiers work into a load impedance of 4 ohms, something like two 8-ohm speakers in parallel. Some use an impedance of 8 ohms, or two 16-ohm speakers in parallel. If you don't know the correct load impedance, use 16 ohms on your load resistor. If you must miss it, it's better to miss it too high than too low; this type of amplifier can withstand too much impedance, but can be overloaded with too little. So, if you don't know, go high. (If you want to make sure, put the ear plugs in and read the full-volume voltage across the unit speakers. This will give you the actual power it develops in service.)

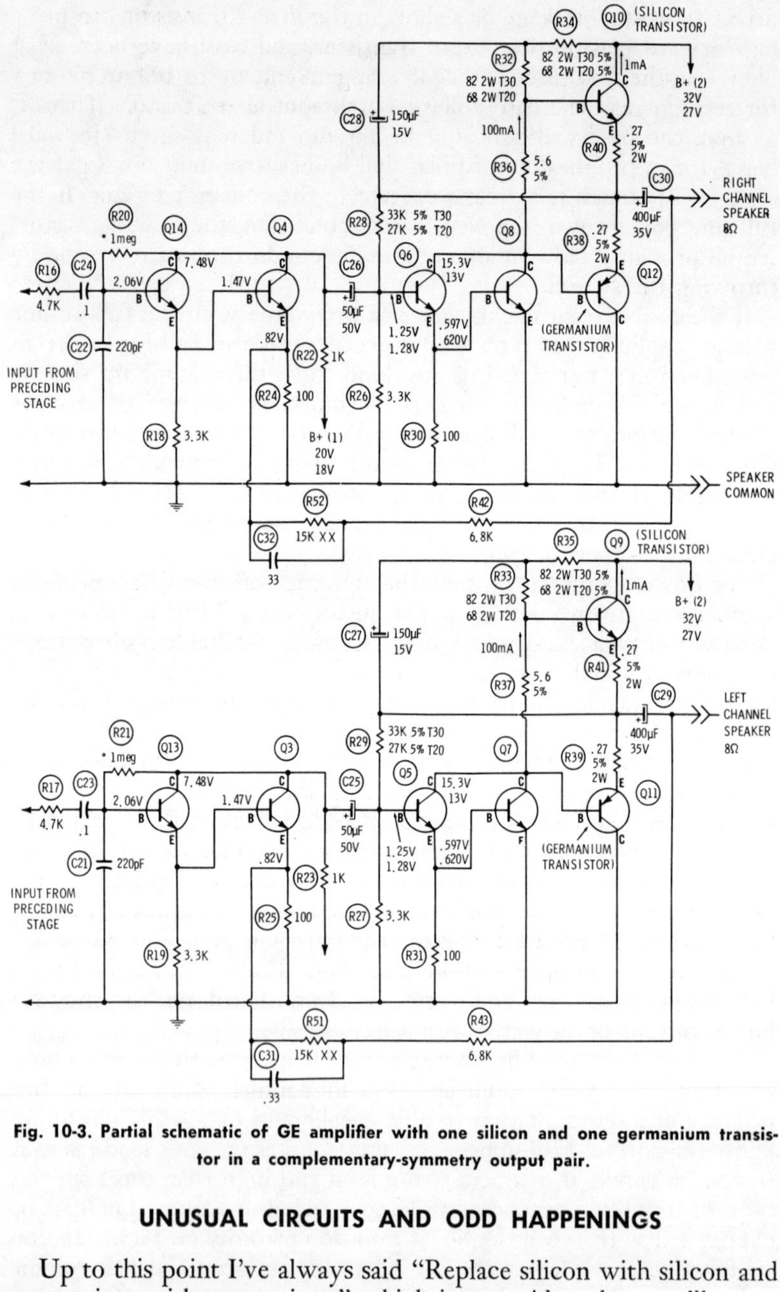

Fig. 10-3. Partial schematic of GE amplifier with one silicon and one germanium transistor in a complementary-symmetry output pair.

UNUSUAL CIRCUITS AND ODD HAPPENINGS

Up to this point I've always said "Replace silicon with silicon and germanium with germanium," which is true. Also, that you'll never find unlike transistors in the same output pair. From all of my research, I was sure that this was universal. Now, I'll have to qualify this, in at least one amplifier circuit that I recently found.

This is a General Electric T20/30 (Fig. 10-3). Although it doesn't *look* different schematically, the npn transistor at the top is a silicon and the pnp at the bottom is a germanium. This, of course, is all right *if* the amplifier is *designed* this way. Biases must be correct for each one, but this is a simple matter. With correct bias, there's nothing to keep silicon and germanium transistors from getting along with each other. However, in amplifier circuits originally designed for the same type transistors, always use this type for replacement. I *do* know, from experiment, that one Si and one Ge transistor in a circuit designed for silicons does cause heavy distortion.

The Odd Happening

The odd happening was the result of an accident, but I thought I'd report it for what it was worth. The "subject" was a Magnavox R220 stereo amplifier. One of the output transistors in the left channel had shorted. Other than the type and voltages, no service data were available at the moment, and this was a rush job. So, I selected a complementary-symmetry pair of silicon transistors, HEP-704/HEP-705 which looked as if they'd fill the specs. They did.

The rest current was pretty low, though. I had made the current tests on the right channel, and noted about 20 mA rest current; now I had only about 3 or 4 mA. However, full-load currents were about the same, 600 to 650 mA peak. Volume was good, and the tone was excellent (ear tests on radio signal).

However, that difference in rest currents bothered me, so I rechecked. Taking the right channel output transistors out and checking them revealed that they were obviously *germanium* transistors. The telltale leakage of about 75 or 80 microamperes, with good beta, showed this up at once. The silicons, of course, showed no leakage.

I had silicons in a circuit designed for germaniums, and they were working. The scope showed practically no crossover distortion on a sine-wave pattern on either channel. There is, of course, no guarantee that this will work in other amplifiers, but I pass it along for what it is worth.

Index

T

V

W

Z